The Nation's Hangar

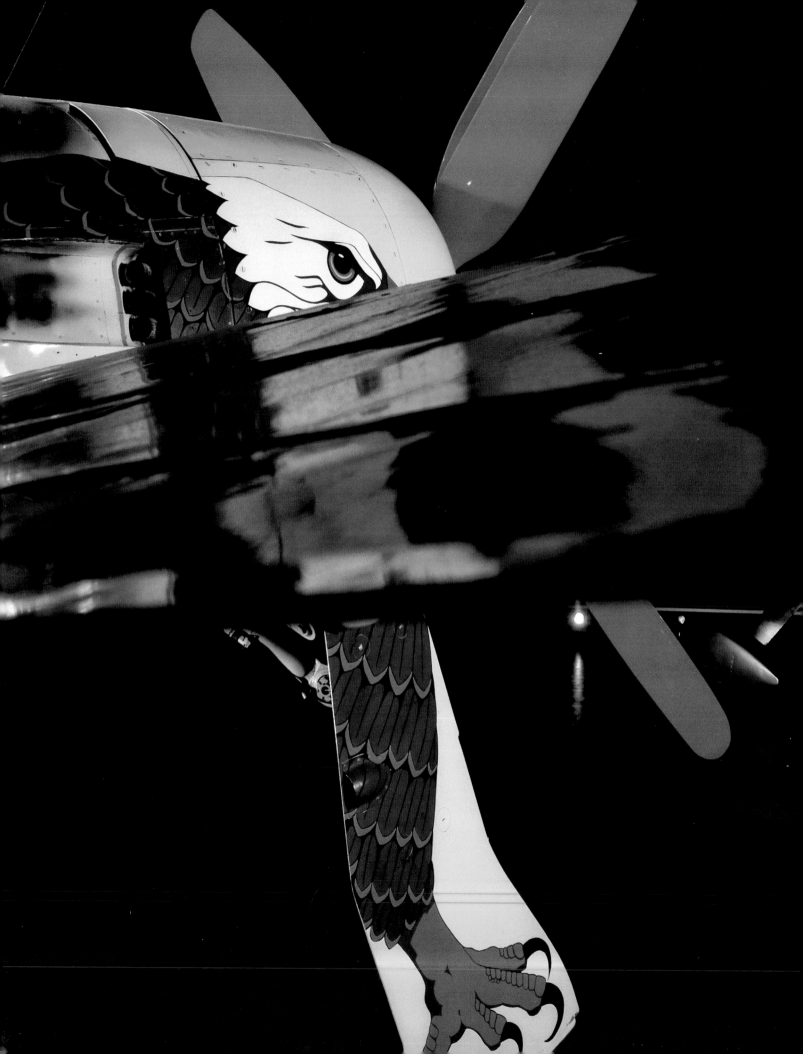

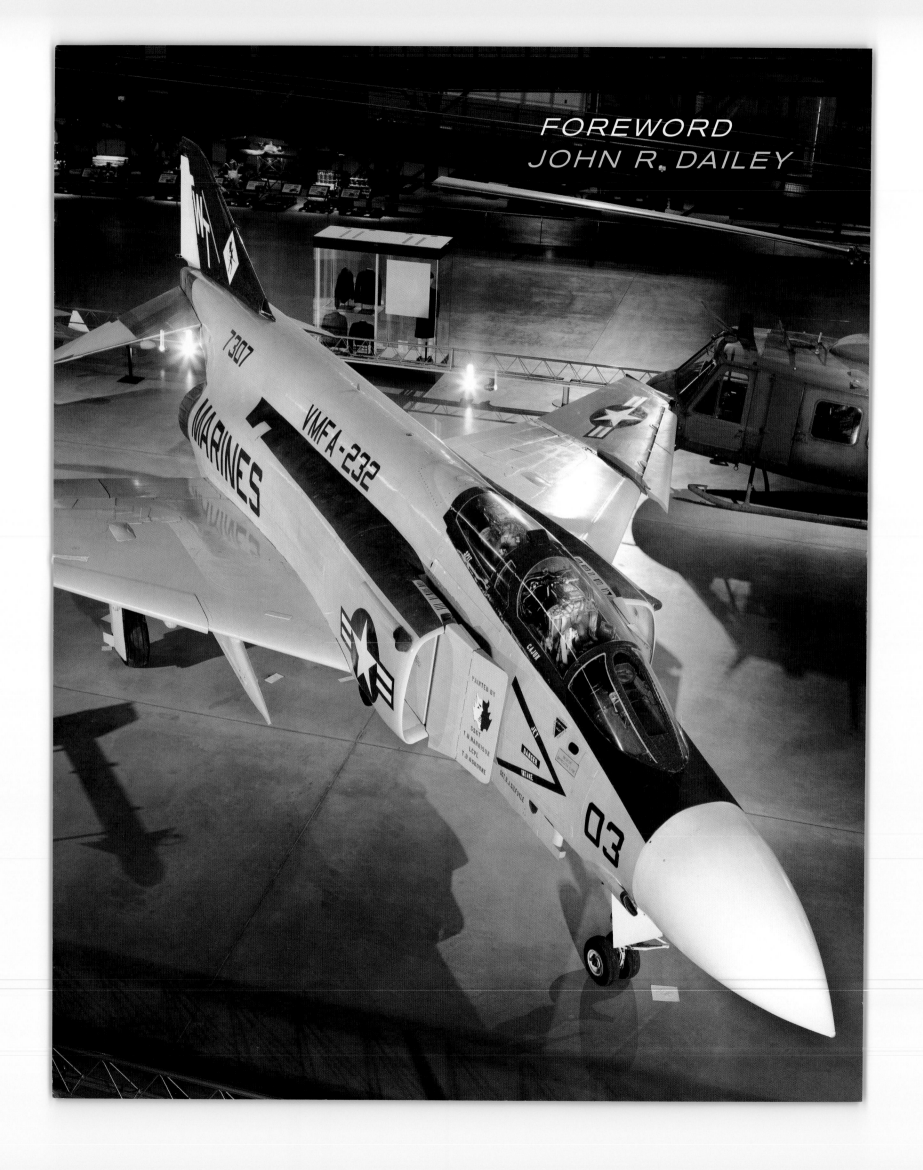

Over a century has passed since Wilbur and Orville Wright took to the air in the first powered, controlled, heavier-than-air flight. Since that cold, windy day on the beach near Kitty Hawk, North Carolina, the airplane has transformed the world in astonishing ways. Originally a miraculous novelty, the airplane quickly became an indispensable tool for commerce, transportation, and war. In 1903 one could only imagine the possibilities of worldwide travel by air; today one cannot imagine traveling around the world in any other way.

The Smithsonian Institution has played a role in this dramatic story. Samuel Pierpont Langley, the third secretary of the Smithsonian, was a contemporary of the Wright brothers. His aerial experiments in the late 19th century paved the way for future pioneers, though his attempts at manned flight failed in 1903. The Smithsonian assisted the Wrights in 1899, and supported rocket pioneer Robert Goddard. In 1915 the Smithsonian helped create the National Advisory Committee on Aeronautics (NACA), NASA's predecessor, which made critical discoveries leading to safer, more efficient flight.

In keeping with the Institution's original mandate for the "increase and diffusion of knowledge," the Smithsonian has led the way in preserving the history and technology of the artifacts of flight. From its first collection of Chinese kites in 1876, the nation's collection of aeronautical artifacts has increased to over 33,000 objects encompassing virtually every aspect of aviation and spaceflight. As the collection expanded, the Smithsonian's ability to display these treasures often failed to keep pace—until now. Following the successful 1976 opening of the National Air and Space Museum on the National Mall in Washington, D.C., momentum grew to construct an extension at a nearby major airport where our large aircraft and spacecraft could be displayed. This dream became a reality on December 15, 2003, with the opening of the Steven F. Udvar-Hazy Center near Washington Dulles International Airport.

In addition to the 70 aircraft we exhibit on the Mall, we now display 200 aircraft and thousands of smaller, equally important artifacts in a magnificent new building over three times larger than the main museum. The aircraft exhibited here cover the entire history of flight and are some of the most significant examples of their types. This modern facility has enabled us to show the vast majority of our superb collection to the public, many objects for the first time. Now, with the completion of the Udvar-Hazy Center, we will have an outstanding preservation, restoration, and storage facility where all of the objects, including our priceless archival holdings, will reside in the latest climate-controlled environments, preserving our nation's aerospace treasures for generations to come. It is indeed the Nation's Hangar.

John R. Dailey
Director
Smithsonian National Air and Space Museum

This McDonnell F-4 Phantom II last flew with the U.S. Marine Corps.

A New Beginning:
Building the Udvar-Hazy Center

CHAPTER 1

he year 2011 marks a significant milestone in the history of the National Air and Space Museum (NASM). Since the opening of the Steven F. Udvar-Hazy Center in 2003, the Museum has actively worked to complete Phase Two of the project, namely the design and construction of a state-of-the-art preservation, restoration, and storage facility. After much hard work, this has come to pass. The facilities completed during Phase Two of the Udvar-Hazy Center construction replace the Paul E. Garber Preservation, Restoration, and Storage Facility as the primary restoration and storage facility for NASM. Named after the Museum's first curator, a visionary who collected most of the NASM's most influential and historically significant pieces, the Garber Facility had served the Museum for 60 years. It was time for a change.

previous spread:
The Lockheed SR-71 dominates this view of the interior of the Stephen F. Udvar-Hazy Center, flanked by the Curtiss P-40 on the left and the Vought F4U Corsair on the right, with the Space Shuttle behind it.

Paul Edward Garber, aircraft collector without peer, almost single-handedly built the world's finest collection of aeronautical artifacts in his 70-year career at the Smithsonian Institution.

Tucked away in Silver Hill, a corner of Prince George's County, Maryland, seven miles from downtown Washington, the Garber Facility is rather nondescript. Surrounded by rundown strip malls and gas stations, a lone Polaris missile sits upright in the parking lot along Old Silver Hill Road, next to the local fire department. Visible behind the perimeter fencing is an unobtrusive sand-colored building with a large blue-and-gold seal of the Smithsonian Institution hanging prominently on the front wall. As inconspicuous as it may be, until the opening of the Udvar-Hazy Center this was the entrance to Valhalla for the aviation enthusiast.

top to bottom:
The Arado Ar 196A-5 was stored outside in Suitland, Maryland, for many years.

Wing components of American and Japanese aircraft await attention while in storage at the Paul E. Garber Facility.

There are 33 buildings at the Garber Facility, shared between NASM, the National Museum of American History, and other support organizations of the Smithsonian. Before the move to the Udvar-Hazy Center, NASM used 23 of these buildings. Eighteen were used for shipping and receiving, dense storage, and support functions, especially as an annex for NASM's Archives Division, and the other five contained thousands of aviation treasures. These buildings were accessible to the public for some time through specially guided tours, but in later years were closed to the public. With the completion of Phases One and Two of the Udvar-Hazy Center, the functions of many of these buildings are combined in one new massive, high-tech structure that features a restoration hangar four times larger than the one at the Garber Facility.

As extensive as the National Aeronautical Collection is, the problem of finding enough space to house the aircraft and spacecraft properly has dominated NASM's planning since the opening of the Museum building on the National Mall in 1976. Still flush with the excitement of opening what would immediately become the world's most popular museum, the curatorial staff was painfully aware of the space limitations placed upon future collecting. The size of the facility at Silver Hill, even after it was refurbished and renamed in Paul Garber's honor, was insufficient to permit future growth.

Barring expansion at the Garber Facility, it would soon be impossible to collect any new military or commercial aircraft, as the size of even the smallest of these aircraft would strain the Museum's ability to house and protect them correctly. For years the Museum had compensated for the lack of space by cobbling together off-site storage solutions and placing pieces of the collection on loan. Some of the larger artifacts, such as the Boeing 307 Stratoliner and the Boeing 367-80, were stored in the Arizona desert. Other large aircraft, such as the Curtiss NC-4, the first aircraft to fly across the Atlantic, were placed on loan to museums around the country. The Space Shuttle *Enterprise* and the Lockheed SR-71 Mach 3 reconnaissance aircraft had to remain behind locked doors in temporary housing at Washington Dulles International Airport. Other aircraft, such as the Boeing B-29 Superfortress *Enola Gay* and the Martin B-26 Marauder medium bomber *Flak Bait*, were left in storage with only the forward cockpits and smaller parts displayed in the Museum. For the Museum to continue to fulfill its mandate to collect and display the most historically and technologically significant aircraft and spacecraft, a long-term solution was clearly required.

What was needed was a large facility with a huge, hangar-like building located at a major airport where future aircraft could simply be flown in and taxied into position for display. This new facility would have to be situated as close to the nation's capital as possible because it was to serve as an annex to the National Air and Space Museum on the Mall, identical in purpose to the Garber Facility only much larger. Furthermore, the new facility would allow the Museum to open the building to the public seven days a week, unlike the Garber Facility, and constitute a museum in and of itself.

During the 1980s, Museum officials began searching for an appropriate location. Naturally, the search focused on Washington Dulles International

Airport and Baltimore Washington International Airport, both major airports capable of handling large aircraft. They are close to major highways, facilitating aircraft movement between downtown and the proposed annex as well as providing ready access to visitors. Both were promising sites with a large surrounding population base. On January 29, 1990, the Smithsonian's Board of Regents selected Dulles as the best solution for the Museum, primarily because of its larger size and the fact that the Space Shuttle *Enterprise* and other aircraft were already stored there. However, this location wasn't finalized until after a long fight played out in the halls of Congress.

Concurrent with the Museum's search for a new annex was the collapse of the Soviet Union and the end of the Cold War. With the communist threat on the wane, U.S. military expenditures were cut, including funding for numerous air bases. Realizing the closure of these military facilities could have a devastating effect on local economies, local politicians and their congressional delegations nationwide pressed their cases for establishing NASM's annex at one of their soon-to-be-defunct bases. While clearly not in the interest of the Museum, which needed an annex close to the main building in Washington, the pressure was intense.

The magnitude of the project blinded many to the facility's original purpose. Intended as a straightforward complex with four basic hangars, the project mushroomed, after the intervention of the Smithsonian's senior leadership, into an expensive multidisciplinary project encompassing several other Smithsonian museums. Soon the annex was being presented as an independent museum to serve the interests of various constituencies. Denver and other metropolitan regions hoped that the presence of a large Smithsonian facility in their city would greatly increase their tax base, create jobs, and promote tourism. The fact that NASM would have to ship its restored aircraft 1,600 miles from Washington to Denver was lost in the smoke of battle.

Eventually, the massive project was scaled back to a form resembling the original idea. Senator Jake Garn of Utah led the voices of reason in persuading Denver and other interested parties to leave the annex in Washington. One cost of the victory was an agreement stipulating that the annex had to be funded primarily from private sources. While Congress would supply seed money and the Commonwealth of Virginia generously agreed to build vital roads, taxiways, and additional infrastructure, the Smithsonian was responsible for raising the bulk of the almost $300 million projected cost.

Owing to the effort of dedicated individuals and officials from federal, state, and local governments, on August 2, 1993, President William Clinton signed into law the bill authorizing the construction of the annex at Washington Dulles International Airport. The then-administrator of the Federal Aviation Administration (FAA), retired U.S. Navy Vice Adm. Donald D. Engen, allocated a large parcel of land to the Museum. Admiral Engen would eventually become director of NASM, and through his untiring efforts greatly advanced the Museum's work toward making the vision of this project a reality.

The site was eventually changed to a new location on the airport's southeast corner; a 40-year lease was signed between the Museum and the Metropolitan Washington Airports Authority (MWAA) on November 12, 1998. Encompassing 176.5 acres, the site borders U.S. Highway 50 on the south and Virginia Highway 28 on the east. The 760,057-square-foot main building is located southwest of runway 01R/19L, just outside of the runway approach. Funding for the construction of a new highway interchange from Route 28 and the entrance road, including ample parking for both passenger cars and buses, was provided by the Commonwealth of Virginia. The state also provided all of the necessary infrastructure and utilities, which included clearing and grading the site. Basic fire and emergency medical support for the facility are provided by the MWAA.

Internationally recognized architectural firm Hellmuth, Obata & Kassabaum (HOK) took the lead in the design of the new building. Known for its soaring yet practical structures, HOK previously had designed the National Air and Space Museum on the Mall, and thus was very familiar with the challenges of aerospace museum design. Working closely with the Museum staff, HOK proposed several different designs intended to meet various funding goals.

The design finally selected was both the most practical and the most expensive. It entailed a building with four separate hangars, each connected to the others to keep the entire complex under cover. Each hangar building had a hangar door at one end and a dedicated exhibit gallery at the other. This massive building mimicked the design of the downtown museum while allowing the addition of new hangars in the future. A phased construction program allowed for the completion of the restoration shop, archives, and storage facilities as time and money permitted.

While planning progressed, the determined efforts to raise the $300 million necessary to build the complex had met with mixed results until a generous benefactor stepped in and made the Museum's dream a reality. Funding had slowly accumulated through the years but the amount was clearly inadequate to begin construction. A massive consolidation of the defense and aerospace industries following the end of the Cold War eliminated many prime sources of funding and generally strapped those that remained. Tight budgets meant that few dollars remained for philanthropy.

For several years the Museum had been talking with the International Lease Finance Corporation (ILFC), the world's largest commercial aircraft leasing company. In a meeting with Museum officials in mid-1999, ILFC's president and chief executive officer, Mr. Steven F. Udvar-Hazy, agreed to donate a substantial sum on behalf of his company. While Museum officials were quite pleased, they were unprepared for his next statement. A refugee from the Hungarian Revolution of 1956, Mr. Udvar-Hazy had found success in his adopted country and wished to repay the nation for providing him with freedom and opportunity. Over lunch, he offered to donate an astounding $60 million of his personal fortune for the construction of the new facility. With this selfless gift—the largest gift to the Smithsonian at the time—the Museum was able to clear its financial hurdles and authorize construction.

Mr. Steven F. Udvar-Hazy (left) stands next to General J.R. "Jack" Dailey (USMC, Ret.), the director of the National Air and Space Museum.

The Smithsonian Board of Regents unanimously voted to name the new facility after Mr. Udvar-Hazy.

After a careful search, the Hensel Phelps Construction Company of Greeley, Colorado, was selected to build the first phase of the Udvar-Hazy Center. In a happy ceremony on October 25, 2000, the first earth was turned. Armed with chromed shovels were William H. Rehnquist, Chief Justice of the United States and Chancellor of the Smithsonian Institution, Mr. Steven F. Udvar-Hazy and his wife Christine, Smithsonian Secretary Lawrence Small, NASM director, General J.R. "Jack" Dailey, USMC (Ret.), and Virginia's Lieutenant Governor John R. Hagar.

Construction began on April 10, 2001, with little time to meet a scheduled opening in time for the Centennial of Flight celebrations in December 2003. For the next two and half years more than 500 workers each day from Hensel Phelps and its 80 subcontractors and vendors removed an estimated 30,000 cubic yards of earth, poured 40,000 cubic yards of concrete, and erected 6,500 tons of steel. They built 122 caissons, each 42 inches in diameter, to support the structure, assembled 87,000 square feet of masonry, put in 209,000 square feet of metal paneling, and installed 12 miles of Walker ducting for electrical and communications cabling. By February 2002, the first of the massive trusses for the aviation hangar was erected. The hangar roof was in place by September 2002, followed quickly by the huge hangar doors. Thick concrete flooring was also poured during this time, as construction workers hurried to complete a myriad of tasks.

Concurrent with the hangar construction was the assembly of the Donald E. Engen Tower. Named in honor of Admiral Engen, the late director of NASM and former head of the FAA who had worked so tirelessly toward this goal before his untimely death in 1999, the 160-foot-high tower was begun in the fall of 2001; its metal cab structure was completed by late 2002. The tower houses a large observation deck for visitors to view the workings of Washington Dulles International Airport and learn about the complexities of air traffic control in exhibits provided by the FAA.

Thanks to a generous contribution by the James McDonnell Foundation, construction of the James S. McDonnell Space Hangar commenced in March 2002. This fortuitous gift has allowed the Museum to display its unique collection of space artifacts, featuring the Space Shuttle *Enterprise*, the atmospheric test vehicle for the space shuttle program. NASM is slated to acquire a flown shuttle when the program comes to its conclusion.

Once the Museum took possession of the Udvar-Hazy Center from the builder in early 2003, the collections management staff faced the daunting task of moving the aircraft, spacecraft, and other artifacts. The bulk of these artifacts were moved miles over open road from the Garber Facility to the Udvar-Hazy Center. The staff had been cleaning, repairing, and preparing the aircraft for the move for several years. All of the aircraft were prepositioned and then delivered in a carefully worked out sequence to minimize delivery and installation problems. They were transported by large tractor-trailers, usually at night to minimize traffic delays.

The proximity of the Udvar-Hazy Center to Dulles International Airport proved its usefulness when moving other pieces of the collection into place. For years the Museum had stored several aircraft, as well as the Space Shuttle *Enterprise*, in temporary buildings at Dulles; these were moved to the Udvar-Hazy Center and installed with relative ease. The Museum's Grumman A-6 Intruder Navy attack jet was retrieved from storage at nearby Andrews Air Force Base.

Other large aircraft were flown to Dulles and briefly kept outside until their turn came for installation. For more than a decade, the Boeing Company had generously restored two historic aircraft on behalf of the Museum—the 307 Stratoliner and the 367-80, also known as the "Dash-80." Because of the great size of these two aircraft, the restoration contracts with Boeing stated that both aircraft were to be restored to flying condition because overland transport would be too difficult. After many years of dedicated work by employees and volunteers, both aircraft were readied for their final flights east from Boeing's facilities in Seattle, Washington. Although Boeing always knew that the aircraft were to join the rest of the National Aeronautical Collection at the Udvar-Hazy Center, more than a few tears were shed when they lifted off for Dulles.

The striking architecture of the Udvar-Hazy Center is immediately apparent to visitors.

In 1989, Air France signed a letter of agreement to donate a graceful BAC/Aerospatiale Concorde supersonic transport to the National Air and Space Museum upon the aircraft's retirement. On June 12, 2003, Air France honored that agreement by donating Concorde F-BVFA(205) to the Museum upon completion of its last flight, flying it in directly to the Udvar-Hazy Center.

While staff loaded and unloaded the aircraft, other workers moved the aircraft into position using cranes, forklifts, tractors, and elbow grease. The responsible curators, the chief designer, and the collections management team had worked out the precise placement of the aircraft ahead of time to ensure that the aircraft fit into the building, were displayed attractively, and were exhibited thematically. On the surface a simple task, in practice it was far more difficult. However, through the use of computer imaging and great attention to detail by the chief designer, William "Jake" Jacobs, the job was accomplished in time.

The Boeing B-29 Superfortress *Enola Gay* was the first aircraft moved into the Udvar-Hazy Center. The centerpiece of the new building, the *Enola Gay* required special treatment because of its size, central location, and unique display. The Museum's restoration staff shipped the famous bomber to Dulles in sections and reassembled the massive aircraft on the floor. Special stands were fabricated to present the aircraft eight feet off the ground to allow visitors safe access to the underside of the aircraft and its bomb bay. Other World War II-vintage aircraft were positioned underneath the *Enola Gay* as they were completed to maximize the existing floor space.

While all of the largest and heaviest aircraft are displayed on the floor, the rest of the aircraft are suspended in two levels from the massive roof structure, which like the main NASM building on the National Mall was designed specifically for this purpose. Each of the trusses is designed to support up to 20,000 pounds equally along its span, which is the equivalent of the empty weight of two World War II-vintage fighters. Cables from specially designed attachment points on the triangular trusses suspend the aircraft. In order to maximize the viewing potential of each artifact, the aircraft are suspended either 25 feet or 42 feet above the floor.

The first phase of construction, which encompassed the main aircraft and spacecraft hangars, the Engen Tower, an IMAX® theater, restaurants, a gift shop, and offices, opened to the public on December 15, 2003, as part of the 100th anniversary celebration of the Wright brothers' first powered flight.

The Udvar-Hazy Center's location at Washington Dulles International Airport makes possible direct delivery flights of aircraft to the National Air and Space Museum. Here, a tractor tows the Concorde supersonic transport into position while the Boeing 307 Stratoliner, the world's first pressurized airliner, waits its turn.

Historically significant aircraft such as the Boeing B-29 Enola Gay, which dropped the first atomic bomb (top), the black Northrop P-61 Black Widow night fighter (center), and the Northrop N1M flying wing (bottom) grace the Boeing Aviation Hangar at the Udvar-Hazy Center.

Flush with success and anxious to increase the preservation, restoration, and storage capabilities of the Museum, NASM staff began taking steps to get the second phase of construction off the ground. NASM's development office worked closely with the Museum's board to identify and secure the $55 million of donations necessary to build the new wing, which came to be known as "Phase Two." In a show of support for the Museum's mission to promote aerospace education, much of the money came from the generosity of board members themselves. In 2008, Airbus Americas, Inc., gave $6 million, enough funding for the Museum to begin building.

Architects Hellmuth, Obata & Kassabaum were called upon to design Phase Two, tasked with creating an addition to the Udvar-Hazy Center that would meet the Museum's restoration, preservation, and storage needs. The proposed design, which was heartily embraced, consisted of a large rectangular structure built alongside the southwest wall of the Boeing Aviation Hangar, immediately south of the James S. McDonnell Space Hangar. While a separate building, the area would be joined directly to the original building through a series of internal, interconnecting walkways, several of which would lead to a dramatic overlook of the restoration shop. The conservation laboratory, Archives facilities, and deep storage were put on the southernmost end of the structure, with office facilities for Museum staff scattered throughout. The construction company Hensel Phelps was selected to execute the design; having worked with Hellmuth, Obata & Kassabaum on Phase One of the project, things went smoothly. The construction of Phase Two was completed in 2011.

The Mary Baker Engen Restoration Hangar provides four times more floor space than the restoration shop at the Garber Facility. Shown on the floor is the fuselage of the Sikorsky JRS flying boat, a survivor of the attack on Pearl Harbor in 1941.

The Mary Baker Engen Restoration Hangar, named in honor of Mary Engen—Admiral Engen's widow and previous board member—was made possible by a $15 million donation from D. Travis and Anne Engen. The Restoration Hangar was designed to provide a modern, environmentally controlled workspace for the restoration, preservation, and treatment of aircraft and spacecraft while maximizing public education opportunities. Glass-enclosed and visible to the public from catwalks that connect the main building to the new wing, visitors to the Restoration Hangar can watch restoration technicians at work while learning from accompanying exhibits about the artifacts and restoration techniques they are seeing.

The Mary Baker Engen Restoration Hangar is four times the size of the restoration space in the Garber Facility, allowing the Museum to restore more aircraft and spacecraft at a time than was possible before. The new space also lets technicians keep the larger aircraft in the collection in once piece during restoration; at the Garber Facility they could only be restored piecemeal.

Important work also takes place behind the scenes, in the Museum's Archives, a state-of-the-art research facility where up to 16 researchers at a time are able to access the 12,000 cubic feet of original documents, 1.75 million photographs, and 14,000 films and videos. Among the Archives Division's holdings is a huge collection of structural drawings for hundreds of different aircraft. This invaluable repository provides priceless assistance to the curatorial and restoration staff for preserving NASM's aircraft, as well as information for restoration of aircraft in private hands. Moreover, the Archives Division maintains the corporate and personal papers of significant companies and individuals.

Although they are the most visible artifacts in the collection, there is far more to the National Air and Space Museum than aircraft and engines. Much of the Museum's collection of spacesuits, full- and partial-pressure suits, leather flight jackets, military, commercial, and civilian uniforms, and other memorabilia are on display and in deep storage at the Udvar-Hazy Center. Artifacts on display are housed in open storage and specially designed thematic display cases and exhibit stations.

Most of the collections not on display are now kept in well-protected, climate-controlled rooms under the watchful care of the Collections Processing Unit staff of professionals. Special storage units hold unique objects in ideal environments based on the needs of the objects. Compact shelving maximizes the available storage space so that the Museum's most valuable artifacts will remain preserved for future generations to enjoy.

Named for the Emil Buehler Foundation, which was the primary donor for this facility, the Emil Buehler Conservation Laboratory is where highly trained conservators examine, treat, and preserve smaller artifacts for exhibit and storage, and assist the restoration technicians in assessing the preservation needs of larger artifacts such as our aircraft.

The successful completion of Phases One and Two of the Udvar-Hazy Center finds the National Air and Space Museum in a better position than ever to fulfill the Smithsonian's mission to increase and diffuse knowledge. The new facilities offer an improved capacity for ongoing historical and technical research in aerospace history and technology and strengthen the collection overall through increased preservation, restoration, and storage capabilities.

As completed, Phase Two, seen below, is immediately south of the James S. McDonnell Space Hangar.

These facilities also broaden access to the Smithsonian's collections and improve the visitor experience, supporting NASM's goal to educate and inspire. Thanks to the new buildings, more of NASM's vast 60,000-object collection is on view than ever before and will be available to the public for decades to come.

Restoration
and Conservation

CHAPTER 2

n the new **Mary Baker Engen Restoration Hangar** building at the Steven F. Udvar-Hazy Center, several artifacts at a time can be worked upon by the Museum staff of expert specialists, most of whom are certified airframe and power plant mechanics. Many learned their trades in the armed forces before joining the Smithsonian. While each knows the intricacies of working on contemporary aircraft, all have acquired additional skills and expertise from their on-the-job experience. No amount of formal training can teach all of the minutiae of working on a rare World War I wood-and-fabric biplane or an exotic Japanese floatplane bomber. These skills are acquired through years of experience and passed down to new technicians who join the staff when their elders retire. Far more than technicians, these men and women are true craftspeople, specializing in fabric, wood-working, painting, metal working, or a host of other skills. The results are impressive. With painstaking effort, NASM's conservation staff has saved dozens of aircraft from the ravages of time and deterioration. Their task is a difficult one, eagerly carried out in a spirit of mutual respect, cooperation, and refined specialization.

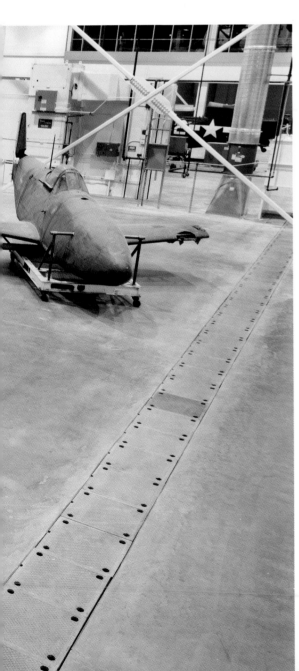

An artifact enters the National Aeronautical Collection after careful consideration by the appropriate curatorial departments. The Aeronautics Division is responsible for deciding which aircraft are brought into the Museum, which ones are placed on loan, which ones go on exhibit, and which ones are removed from the collection. A careful system has evolved to consider every issue before the status of an aircraft is changed.

The acquisition of an aircraft begins with the appropriate curator among the eight currently in the Aeronautics Division. He or she weighs the merits of the acquisition, determines whether it will enhance the collection, and, if so, presents it to the Aeronautics Division's collections committee. The committee, which consists of all eight curators plus the Division's chairman, discusses the proposed acquisition based on the guidelines of the Aircraft Collections rationale, a detailed document that outlines the aircraft-collecting plan of the Museum.

Three criteria are used to determine whether or not an aircraft is suitable: historical, technological, and practical. The historical criterion is divided into five subcategories:

1. The aircraft is recognized as a milestone in aviation history, such as the 1903 Wright Flyer, the world's first aircraft.

2. It represents an important, well-defined era or event in the evolution of aviation, such as Charles Lindbergh's Ryan NYP, the *Spirit of St. Louis*, in which he flew solo across the Atlantic in 1927.

3. It represents political or social and cultural factors that have affected aeronautical history, such as the Concorde supersonic airliner.

4. It has had a continuing long-term effect on society and culture, such as the Boeing B-29 *Enola Gay*, which dropped the first nuclear bomb.

5. It has played a significant short-term role in shaping the course or outcome of a major historical event or era, such as many World War II aircraft

There are four technological criteria:

1. The aircraft represents a significant advance in aircraft performance, for example the experimental Bell X-1 that first broke the sound barrier.

2. It represents a significant advance in the application of technology to a specific aeronautical role, for example the classic Douglas DC-3 airliner.

3. It represents the development of aeronautical technology that has since been widely applied in other fields, for example the North American X-15 hypersonic research aircraft.

4. It represents political and economic factors that influenced the development and application of technology, for example many post-World War II military and civilian aircraft types.

Finally, there are four practical criteria that must also be considered:

1. The aircraft can be obtained, preserved, restored, and exhibited at a reasonable cost.

2. It can be transported to the Museum or to and from temporary storage.

3. It has research, scholarly, and/or educational value.

4. It meets the physical requirements for exhibition.

If approved by the Aeronautics Division's committee, the responsible curator makes a formal presentation to the Museum's official Collections Committee, which is comprised of the chairs of the curatorial departments, two representative curators, the Conservator, the Chief of Collections Management, and other interested parties. This committee weighs the same selection criteria and votes on whether or not to acquire the aircraft. If approved, the committee's recommendation is forwarded up the chain of command to the director, who ultimately makes the final decision.

Once an aircraft enters the collection, it can no longer be treated as it was when it was in service; it is now an artifact. In the past, the most widely accepted technique was to restore an aircraft to its original factory-delivery condition. This produced beautiful aircraft in resplendent paint but often at the cost of originality. A thorough restoration that replaces numerous parts and strips the paint from the aircraft destroys much of the originality of the artifact. In some cases, for example when the aircraft is in dire condition, this is acceptable, but the desire today is to preserve the artifact as much as possible, for in this way the integrity of the original technology and history can be maintained. Ideally, when new aircraft are acquired, they will be placed in a proper physical environment that will forestall the effects of time and decay. As with any other artifact, objects must be handled as little as possible, whether it is a ruggedly-built aircraft or a fragile work of art. When that is not possible, the Museum must act.

An extensive archival materials library and a large inventory of original components help ensure accurate preservation or restoration of aviation artifacts, a primary goal of the Museum.

The Museum's Collections Management Policy details explicitly the Museum's requirements and responsibilities regarding its artifacts. Preservation is foremost:

> NASM is committed to the preservation of all its collections through (1) an active preventive conservation program, (2) the preservation, restoration, and treatment of specific artifacts, employing conservation principles, (3) the responsible exhibition, movement, cleaning, and handling of objects, (4) the provision of quality storage environments, and (5) complete documentation of object condition and treatment. Artifact treatments follow the American Institute for Conservation's Code of Ethics and Standards of Practice to the degree practical.

The restoration process, when required, must also follow established policy. While the effort is always collegial, the roles and responsibilities of the individuals involved are clearly outlined and must follow established methods:

> Artifact restorations are always historically accurate and reflect the technology original to that artifact. It is the combined responsibility of the curator (who takes the lead and has final decision-making authority), restoration foreman, restoration specialists, and conservator, to determine (1) the goals of a specific restoration, (2) the object's final configuration, and (3) methods of treatment, with an

At work on the starboard wing of the Japanese Aichi Seiran bomber.

emphasis on reversibility. Original components and materials are always used when available; they are preserved or returned to their original configuration. Repairs are made if necessary and marked as such. Complete documentation, including restoration log books, is maintained in the curatorial files; a summary of treatment given an object is filed in the appropriate accession file.

How an aircraft will appear after it is restored is also carefully thought out. Unless it is impossible to do so because the history of the particular aircraft is not known and the original markings have been removed, all aircraft must be restored to represent the event for which, or the operating regime in which, they are most well known. The re-marking of the aircraft must be as faithful to the original markings as practical. When presented with a choice, such as a U.S. Navy aircraft that carried several different variations of an approved paint scheme, the choice must be to present the aircraft in its most important scheme. For instance, if the aircraft has a significant combat history it should be painted in the markings in which it fought and not those in which it was delivered before seeing combat. There are cases when an aircraft is a generic type with an insignificant history. In this event, it is appropriate to paint the aircraft in a scheme that best represents the type while in service. It is not appropriate in this case to paint it in the markings of a specific, well-known example, such as the markings of a similar aircraft flown by an "ace." That would be deceptive and misleading. Because the process is irreversible and damages the aircraft, the earlier policy of mounting a detailed identification plaque on the inside of restored aircraft is no longer appropriate.

When the project is completed, the restored aircraft is rarely flown. According to the Collections Management Policy:

> While NASM's artifacts may be returned to near-flight condition, they are rarely flown, and engines rarely run. NASM will not risk objects in the National Collections in flight demonstrations. Aircraft will only be flown if, in the judgment of the director, flight is the most practical method of movement.

These are the principles of artifact preservation and restoration that have guided the staff for many years and will continue to do so for the foreseeable future.

Aircraft are selected for restoration based on the condition of the aircraft and exhibit requirements. Some aircraft are in better shape than others. Those with the most pressing need for conservation are given preference because any artifact, whether made of metal or of wood, has a limited life, as the forces of decay are constantly and inexorably on the attack. The job of the Museum is to prolong that life for as long as possible, preferably for hundreds of years.

The Museum has developed a system to identify the level of treatment required by all the aircraft in the collection. Three different levels were determined and are applied. Level 1 requires the least care. These artifacts are

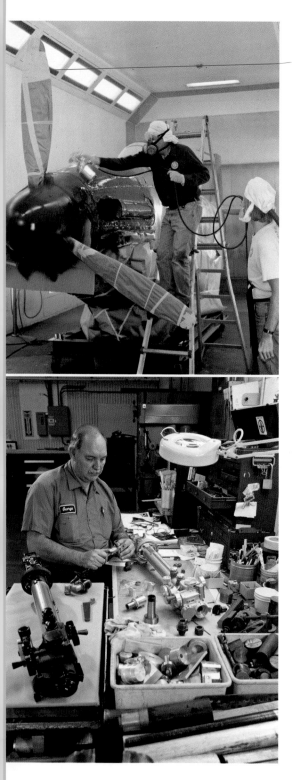

Restoration and conservation staff at the Garber Facility at work on various aircraft components.

either already restored or have been well preserved since their acquisition and need minimum care for exhibit. Those in Level 2 are in generally good shape as well, and require only minor work such as painting or repair before they are presentable. Those in Level 3, however, are quite different. These artifacts require a thorough treatment involving major disassembly to search for and arrest corrosion and to repair significant damage. Only then would a Level 3 artifact be suitable for display.

One other important consideration is weighed before an aircraft is chosen for restoration. Because objects are brought into the collection so that they may be shared with the public, aircraft and other artifacts must eventually be prepared for display. When the Museum decides to create a new exhibit, the appropriate curator assumes the responsibility of selecting the artifacts and preparing the label script. If the artifacts require preservation or restoration treatment, they are placed ahead of other objects that may be in need of more thorough care but are not immediately required for display. Thus, the selection of artifacts for restoration is based on exhibit requirements as well as preservation needs.

To address these questions, members of the curatorial and collections management staff meet periodically to determine restoration priorities. The first step is for the curator to prepare a curatorial package that outlines the extent of the restoration and the appropriate final configuration of, for instance, an aircraft. These guidelines would also include a thorough history of the individual aircraft type and the specific history of the aircraft to be restored. In cooperation with the restoration staff, the appropriate restoration techniques are outlined to serve as a guide. The expertise of the craftsmen is essential, as they best know the correct techniques and materials suitable for the restoration procedures. Once these decisions have been made, the appropriate artifact is moved on to the shop floor and work begins.

The primary concern of the restoration specialist is to assess the condition of the artifact and address the extent to which time and exposure to the elements has caused deterioration. Intergranular corrosion of the aluminum alloys is a common problem, particularly with Japanese aircraft. As the composition of the alloy ages, it is gradually weakened, causing the metal to decompose from within. This is revealed when the aluminum alloy delaminates and flakes off. Surface corrosion occurs when the thin protective layer of pure aluminum is violated by a dent or scratch that exposes the more vulnerable alloy to the air. This often results in pitting which, if not too severe, can be treated with a simple protective coating. The repeated expansion and contraction of metal during the rigors of flight, such as wing flexing or repeated fuselage pressurization, can cause stress corrosion. Over time this weakens the metal and allows corrosion to enter along points of stress, particularly around rivets. The corrosion of steel—rust—is also a serious concern around landing gear and engine mounts.

Deterioration of organic materials such as wood and fabric is also a significant problem. When subjected to humid climates, these materials absorb water and become a breeding ground for fungus. The resulting rot destroys

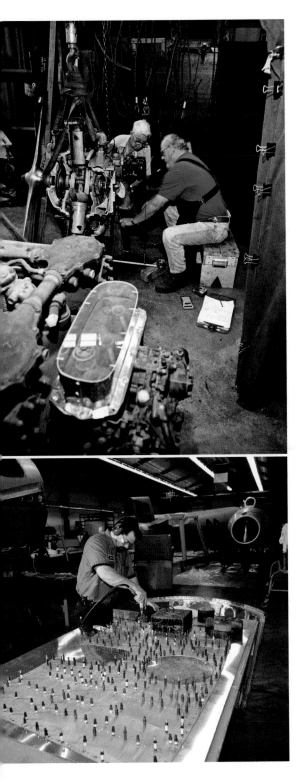

Restoration technicians are expert in fabric, metal, and wood fabrication.

the cellulose that provides wood's strength. Fungicides can eliminate the problem temporarily but the only long-term solution is to place these aircraft in a dry, climate-controlled environment to prevent a recurrence. Fortunately, if it cannot be salvaged, wood is easily repaired, using standard approved procedures. Only as a last resort will replica parts be fabricated. Then, as with metal parts, the object will be permanently marked as a replacement to prevent confusion and ensure authenticity in the future.

Similarly, fabric, whether cotton or linen, is subject to degradation over time. Humidity and the resulting fungal growth can weaken fabric, but the worst culprit is light, particularly ultraviolet light, which can destroy the strength of a fabric. Currently, restoring fabric is difficult, if not impossible. If small repairs are no longer practical, as a last resort the Museum is forced to make the difficult decision to replace the fabric, as it did with the 1903 Wright Flyer and the Spad XIII. In each case, the fabric was severely damaged by light and humidity. Great care is then taken to find modern fabric that is as correct as possible in type, warp, woof, and thread count. The fabric is then remounted using the same stitching techniques as found on the original, and then treated with the appropriate dope at least twice to tighten the fabric Subsequent coats are used to apply the correct color. Antifungal chemicals are also applied.

The old fabric is not disposed of because it is an important part of the aircraft's history and can serve as a model for replicating similar fabric in future restorations. It also preserves the original markings and stitching, which are important for future historians and restorers to study. Therefore, the original fabric is carefully mounted on acid-free backing and placed in environmentally-friendly storage for long-term preservation.

Other materials pose unique problems of their own. Since the Second World War, more and more parts have been made with plastics, particularly Plexiglas aircraft canopies. Plastics are subject to the development of minute cracks called "crazing" and can also develop clouding, turning opaque. Careful polishing can reduce the severity of these problems but unfortunately cannot eliminate them. Rubber is similarly difficult to work with; both ozone in the atmosphere and ultraviolet light cause rubber to dry and harden. The opposite can also happen: Rubber can revert to a liquid and flow slowly as it permanently deforms. Once that happens, nothing can be done to restore it. This is true for tires, hoses, and other fittings. Long-term preservation problems presented by the exotic new materials being used increasingly in aircraft and particularly in spacecraft are yet unknown. In this case, proper preservation in a benign environment is the only practical solution.

When preservation or restoration cannot replace a damaged or missing part, the Museum must attempt to locate its suitable replacement. Through the many personal and professional contacts of all involved staff members, as well as advertisements placed in appropriate trade journals and thorough internet searches, even the most obscure part can often be located. As a last resort, the highly trained specialists at the Garber Facility are able to fabricate virtually any part that is required.

A British World War II Hawker Hurricane sits next to a French Nieuport 28 from World War I, both restoration works in progress at the now-closed Garber Facility restoration shop.

The complex treatment given to the famous Boeing B-29A *Enola Gay* well illustrates many of these complicated issues. On August 6, 1945, this massive, long-range strategic bomber dropped the first atomic weapon, destroying the Japanese city of Hiroshima and thus hastening the end of the Second World War. It entered the NASM collection on July 3, 1949, when Colonel Paul Tibbetts, commander of the 509th Composite Group and pilot of the *Enola Gay*, flew it to Orchard Place Army Air Field near Chicago. When the Smithsonian was evicted from its Park Ridge location, the Air Force flew it first to Pyote Air Force Base, Texas, early in 1952 and eventually to Andrews Air Force Base near Washington, D.C., on December 3, 1953. It remained at Andrews until 1961 when space was finally made available at Silver Hill, just a few miles down the road.

Despite its historical importance, the *Enola Gay* languished until December 5, 1984, because of its size. The mighty bomber was then carefully retrieved from storage for its restoration. It is the largest restoration project ever undertaken by the Museum. Large though the restoration shop may have been, it was far too small for a B-29. This single artifact most clearly demonstrates the need for the new restoration shop that will eventually be built at the Udvar-Hazy Center.

Already in parts, the *Enola Gay* was moved through the shop piecemeal, with the forward fuselage and cockpit receiving priority. Though the aircraft had been stored outside for much of its life, it had fortunately not suffered too severely from exposure. The forward fuselage was in remarkably good condition, although some instruments were missing.

To determine the level of corrosion that needed to be treated, the aircraft was first cleaned. In the past, most of the aircraft would have been disassembled, treated, and rebuilt, but because of the *Enola Gay*'s large size and improved restoration techniques, the aircraft was disassembled only as far as necessary. Careful examination, using borescopes and other instruments, allowed the technicians to perform equally effective but less invasive procedures.

To prevent damaging the artifact while searching for corrosion, the disassembly was limited to the wings, control surfaces, engines, and turrets. The wing proved a particular challenge because, in addition to the nests and debris left by birds and other animals, it was attached directly through the fuselage. Removing the carry-through wing required the staff to take off the surrounding sheet metal and a series of massive, tapered bolts, a difficult and time-consuming task. Inside the fuselage, the quilted insulation was found to be in tatters from the ravages of mice and was not salvageable. Instead, the craftsmen fabricated new linings and padding using the old material as a pattern.

Expert knowledge and craftsmanship are required for working with aircraft materials. Every effort is made to preserve originality.

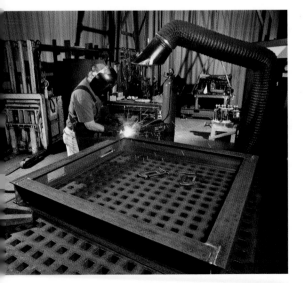

Consulting technical manuals borrowed from the Museum's extensive archival collection, NASM's craftsmen carefully inspected and treated every section of the massive bomber. The manuals were especially helpful in locating the proper location and direction of the B-29's intricate wiring and plumbing systems. Some parts of the aircraft, such as the astrocompass, were in such poor condition that a better example was retrieved from the Museum's large parts inventory and installed in place of the damaged article.

During reconstruction of the fuselage and wings, mechanics worked on the aircraft's powerful Wright R-3350 engines. Two of the engines were refurbished on the shop floor of Building 10. All told, the responsible craftsman spent more than 6,000 hours over many years on these two engines. Expert technicians at the San Diego Aerospace Museum restored the other two. NASM has often allowed other qualified museums to perform restorations, on condition that they meet our exacting standards. This is ensured through a restoration contract that all participating museums must honor. The results are worth the effort. San Diego delivered two immaculately restored engines.

Every available personnel resource was used to restore the *Enola Gay*. In addition to highly trained staff members from NASM and the San Diego

light, Nieuport 28s saw widespread service for this use in every air force. Aircraft such as the de Havilland D.H.4, the Breguet 14, and the German L.V.G. series of biplanes perfected the tactics of close air support by 1918.

In March 1918, the German army achieved a breakthrough on the Western Front, shattering the four-year-long stalemate. Using combined arms and shock tactics, the Germans broke through Allied lines and routed the British and French forces. For the first time, coordinated attacks were made by light bombers and fighters in immediate support of advancing troops. These served as effective rehearsals for the techniques of tactical air power that were so successful in the next world war. With the deadlock broken, at least for a while, the combatant armies were now able to maneuver; this made the airplane even more valuable in locating and attacking the enemy. The breakthrough was stopped only by a combination of the general exhaustion of the German troops, desperate Allied counterattacks, and the timely arrival of American ground forces. The **Halberstadt CL.IV** was typical of the German light bombers used so effectively at this late stage of the war.

The CL.IV was introduced just in time for the March 1918 offensive. These high-performance two-seat bombers were organized into highly effective "battle flights" of four to six aircraft each that struck at decisive points on the battlefield. Each flight was given specific targets to attack at precise times to open a path for the advancing troops at key points along the front. As with many German aircraft, the CL.IV was powered by a 160-horse-power Mercedes engine that gave it a top speed of 103 miles per hour. It was armed with two fixed, forward-firing machine guns and two flexibly-mounted machine guns operated by the observer. Antipersonnel grenades and up to five 22-pound bombs were also carried. By the time the German offensive died out, the Halberstadt and its compatriots were forced on the defensive. In this role it attacked Allied troop assembly points and later flew many nighttime harassment missions. With no armor protection, the CL.IV, like most of its contemporaries, was vulnerable to ground fire but its good maneuverability made it a surprisingly dangerous foe in air-to-air combat.

The Halberstadt in the National Collection is believed to have fought in Belgium during the March Offensive. It was acquired through the generosity of the National Museum of the United States Air Force and beautifully restored by the Museum fur Verkehr und Technik in Berlin.

By 1918, the skies were full of comparable light bombers. A typical French contemporary was the **Spad XVI**. Developed from the Spad XI, the XVI was designed by Louis Béchereau along similar lines to his popular Spad VII and XIII fighters. A sturdy biplane of wood construction and covered in fabric, the Spad XVI was powered by an eight-cylinder, water-cooled, 230-horsepower Lorraine-Dietrich "Vee" engine. This gave the two-seat Spad an excellent top speed of more than 120 miles per hour. It carried two forward-firing fixed machine guns and two rearward-firing flexibly-mounted machine guns.

Though well-built, the Spad XVI, as with the XI, did not handle well and was notoriously heavy on the controls. Worse, the aircraft was tail-heavy and prone to stalling, which often resulted in a spin. As a result, the Spad XVI saw limited use and was quickly withdrawn from frontline units. Of the 235 aircraft built the sole survivor is in the National Aeronautical Collection. On loan to the Air Force Museum, NASM's Spad XVI is very special because it was the personal aircraft used by General William "Billy" Mitchell. The Army Air Service acquired six of these aircraft in August 1918 and this one was assigned to Mitchell, who used it extensively to survey the front. From this aircraft, Mitchell directed the air operations of the Army Air Service during the battles of St. Mihiel, where he led the massive air assault in support of the U.S. ground troops as well as the subsequent Meuse-Argonne Offensive that ultimately resulted in the German capitulation. After the war, Mitchell gave the Prince of Wales an aerial tour of the Rhine from this Spad XVI, after which the aircraft was eventually transferred to the Smithsonian.

Gen. William "Billy" Mitchell's Spad XVI was transferred from the U.S. War Department to the Smithsonian Institution in 1920.

The Golden Age
of Flight

CHAPTER 4

T

the j
unde
set
to er
It wo
this
dent
were
ente
to go
Offic
mod

*The Boeing FB-5 was one of the first
carrier-based fighters in the U.S. Navy.*

*previous spread:
Unique overhead perspective of the
Boeing 307 Stratoliner.*

in the orange-and-blue corporate colors of Gulf Oil and renamed the *Gulfhawk*. With this aircraft, Williams toured the country promoting Gulf aviation products and performing astonishing aerial displays that captivated an entire generation of young enthusiasts. His dive-bombing expertise was witnessed by top surviving World War I German ace Ernst Udet during the National Air Races. So impressed was he with Williams and the accuracy of this new technique that Udet purchased two Hawks and shipped them home. With these aircraft Udet later taught the German Luftwaffe how to dive bomb—a lesson it put to devastating use during the subsequent world war.

During its career, the *Gulfhawk* was re-engined several times, first with a 575-horsepower Bliss Jupiter radial and later with a much larger 710-horsepower Wright R-1820 radial. When the Wright was installed, the long-range tanks were removed and the aircraft rebuilt with an all-metal fuselage. Eventually, when Williams moved on to more modern aircraft, the *Gulfhawk* was sold to a New York aviation trade school. It was later purchased by noted movie stunt pilot Frank Tallman, who rescued it from certain destruction and flew it for many years after an extensive restoration to its current configuration. The *Gulfhawk* was donated to the Museum in 1969.

A contemporary of the Hawk series of fighters was the other Curtiss fighter in our collection, the unique **F9C-2 Sparrowhawk**. This diminutive biplane was originally designed to compete for a Navy requirement for a lightweight fighter. Although the Navy changed the requirement, a special role was found for the Sparrowhawk. During the 1920s and early 1930s, the Chief of the Bureau of Aeronautics, Rear Adm. William A. Moffett, the "father" of naval aviation, had become fascinated with the possible potential use of airships. He envisioned a fleet of dirigibles that would act as aerial cruisers scouting well ahead of the battle fleet. To fulfill his dream, the Navy had already flown several airships and was preparing to launch two new huge dirigibles, the USS *Akron* and USS *Macon*. Earlier tests on the USS *Los Angeles* had demonstrated the practicality of operating fixed-wing aircraft from an airship aloft.

To increase the scouting range of the new airships and to provide for their aerial defense, the Navy purchased eight Sparrowhawks, the first of which was delivered in June 1932. Each aircraft was fitted with a well-braced hook above the top wing, which was designed to latch on to a trapeze mechanism lowered from within the airship. After it was successfully captured, the Sparrowhawk could be hoisted up into an internal hangar bay on the airship. When the F9C-2 Sparrowhawks flew missions during which they were not intended to land, they were operated from the *Macon* without landing gear to lighten the weight and lower the drag. Each Sparrowhawk was built with an all-metal monocoque fuselage with metal-framed wings covered in fabric. A 438-horsepower Wright R-975 provided the power. After the loss of the *Akron* in 1933 and the *Macon* in 1935 with four aircraft on board, the remaining four Sparrowhawks were stripped of their hooks and relegated to utility duties. The Smithsonian acquired the only one that survived the

cutter's torch in 1939. Restored in 1975 and displayed for many years, it was on loan to the National Museum of Naval Aviation in Pensacola, Florida, but has now returned to NASM.

By the mid-1930s, the U.S. Navy was introducing more improved fighter designs as well as dive-bombers and torpedo planes. During this time, the newly formed Grumman Aircraft Engineering Corporation produced the first in a long line of distinguished naval aircraft—the FF-1, an all-metal two-seat biplane fighter with fabric-covered wings. When the "Fifi" first flew in December 1931, it was the first naval fighter ever to exceed 200 miles per hour and the first ever to be equipped with a retractable landing gear. The success of the two-seat FF series invariably led to a request from the Navy for a single-seat version. Consequently, in October 1933 the first F2F flew and, despite a tendency to spin, proved to be a fast and highly capable fighter. Fifty-six of these aircraft were built until they too were replaced by a more powerful sibling, the F3F, which would be the last biplane fighter to serve in the fleet. Like all Grumman fighters, it was exceptionally strong, designed to withstand even the stresses of 9-g maneuvers. This would be an impressive achievement for a modern fighter today, and was an astounding achievement for a 1930s design.

The F3F's remarkable strength, coupled with the design's high maneuverability, made it an ideal choice for demonstration pilots. As before, with the *Gulfhawk*, Gulf Oil purchased a special version of this airplane for Al Williams. Flying under the Grumman company designation of G-22, the **Gulfhawk II** combined an F3F fuselage and 940-horsepower Wright R-1820 engine with the smaller wings of the F2F. The aircraft was stressed for even more stringent maneuvers and was equipped for up to 30 minutes of inverted flight. This combination permitted truly formidable aerobatics, which Al Williams enjoyed with great panache between 1936 and 1948. Having rejoined the service as a Marine Corps reserve officer, Major Williams astonished crowds around the world. In 1938, he traveled with the *Gulfhawk II* to Europe where he performed at air shows in Britain, the Netherlands, France, and Germany. While in Germany, he met up with his old acquaintance Ernst Udet, by now a general in the Luftwaffe. Udet flew the airplane, the only pilot other than Williams to do so. In return, Udet allowed Williams to fly the vaunted new Messerschmitt Bf-109 fighter—a rare privilege.

Our Stratoliner started life as Pan American's *Clipper Flying Cloud*, plying the air routes to South America before it was drafted into transatlantic service. After the war it was discovered derelict at Tucson, Arizona, where it languished until Boeing volunteered to restore the aircraft for NASM. The dedicated crew prepared it and flew it back to its birthplace in Seattle in 1994.

For the next seven years, Boeing employees and volunteers lovingly restored this beautiful aircraft to its original glory. In June 2001, it once again took to the sky resplendent in its highly polished aluminum and blue Pan American livery. The following month it was the star attraction at the Experimental Aircraft Association's annual fly-in at Oshkosh, Wisconsin, where hundreds of thousands of spectators were thrilled by its impressive daily flights and its magnificent restoration. Unfortunately, during a routine test flight, this beautiful aircraft ran out of fuel and was ditched in Elliot Bay near Seattle. Fortunately, the aircraft, while damaged, was quickly recovered from the salt water, drained, and returned to its hangar in Building 2 where its second restoration immediately began. After much dedicated work from hundreds of current and former Boeing employees, the 307 was returned to its glory and flown to its new home in the new Steven F. Udvar-Hazy Center in the summer of 2003.

Golden Age commercial aviation was not always about advanced technology; it was also about practicality. In the early 1930s, when the United States was developing new streamlined all-metal aircraft with stressed-skin, monocoque construction, other manufacturers were relying on older but well-tested methods to create reliable transports. In Germany, the most successful of these efforts was the Junkers Ju 52/3m. Using proven corrugated duralumin metal and a welded tube structure, Junkers had built a long line of successful military and commercial aircraft from as far back as the First World War. The Junkers-F 13 and -W series of single-engined transports won worldwide acceptance. So successful was the Junkers corrugated duralumin that the Ford Company combined it with the high-winged tri-motor layout of the Dutch Fokker transport to create the popular Ford Tri-Motor. Furthermore, Hugo Junkers was the acknowledged inventor of the cantilevered wing, for which he received a patent in 1910.

The awkward appearance of the **Ju 52/3m** belied its reliable performance and was the most successful Junkers transport design. Basically an enlarged three-engined version of his monoplane transports, the Ju 52/3m first flew in April 1932 and could cruise at a surprising 150 miles per hour. It could carry 17 passengers or three tons of cargo and its unique "double wing" ailerons and flaps gave it excellent short-field performance. Because of its versatility, it quickly became Germany's most important military transport after the outbreak of the Second World War. A total of 4,835 were built.

During and after the war, Ateliers Aéronautiques de Colombes of France and Construcciones Aeronáuticas (CASA) of Spain continued to build Ju 52s. In the 1980s, Lufthansa, the national airline of Germany, flew a CASA-built version for publicity purposes. In 1987, the airline generously donated it to

following spread, clockwise from left: Sky and clouds are reflected in the highly polished aluminum of the Boeing 307 Stratoliner.

The main passenger compartments were converted to sleeping berths during flight at night.

A beautifully finished, coral-colored powder room opened to the commode.

Perhaps the most exciting and visible category of aviation during the Golden Age was air racing. The exploits of racing pilots were widely reported and many of their names became household words. In many ways, Jimmy Doolittle, Roscoe Turner, Matty Laird, Lowell Bailes, and many others were as well known then as star athletes are today. The famous Pulitzer Trophy Races, and especially the Schneider Trophy Races of the 1920s between military services and nations, generated much new research and technological innovation, particularly in high-performance engine design, cooling systems, and fuels.

By the 1930s, after government sponsorship ceased because of the Depression, private individuals assumed the role of competitive sponsorship. The great National Air Races held annually in Cleveland, and once in Los Angeles, brought together the finest piloting talent in the nation. The privately designed and built aircraft were essentially over-powered homebuilts that brought much idiosyncratic originality to aircraft design. They were not, however, technologically advanced. Most, if not all, were conventional wood-and-fabric aircraft, often with steel-tube fuselages. They were compact designs designed intuitively for speed. The power was purchased, not developed. The engines were military or commercial designs from the major mainstream manufacturers that were tweaked for more power but not built specifically for racing. Nevertheless, the racing was fun, highly competitive, and excited the imagination of the public by providing a new generation of heroes, even if the industry learned little from it.

Northrop demonstrated the efficiency of the flying-wing design with the N-1M, which maximized lift and minimized drag when compared to conventional aircraft.

Two such aircraft reside at NASM. In 1931, Matty Laird and his chief designer Raoul Hoffman created a compact biplane, the *Super Solution*, to recapture the Thompson Trophy Race they had won the year before. On September 4, 1931, with famous former Army pilot Jimmy Doolittle at the controls, the new Laird *Super Solution* seized the headlines when it crossed the finish line first in the Bendix Trophy cross-country race between Burbank and Cleveland. Inspired, Doolittle kept going and set a new transcontinental mark of 217 miles per hour when he finally stopped at Newark, New Jersey. He thereupon returned to Cleveland to compete in the closed-course Thompson Trophy Race. Unfortunately, problems with his Pratt & Whitney R-985 engine forced him to retire after 7 of the 10 laps were completed. This was the last race for the *Super Solution*. Doolittle later flew the aircraft from Ottawa, to Wichita, and then to Washington, DC, before the aircraft was sold. Doolittle joined forces later with the Granville brothers to fly their famous Gee Bee R series racers. The *Super Solution* was dismantled; in 1948, the Museum received the fuselage while the wings went to the Connecticut Aeronautical Historical Association.

The most famous racer in our collection belonged to the Golden Age's most flamboyant pilot, Roscoe Turner. In 1934, Roscoe was already famous for his flashy military-style uniforms and pet lion Gilmore with whom he briefly flew. He was concerned that, despite his victory in the Thompson Trophy event, his trusty Wedell-Williams racer was falling behind the competition. With little money, Turner approached the Lawrence W. Brown Company of California to build an aircraft of his own design. Working with professors

Roscoe Turner, perhaps the most colorful flyer of the Golden Age, won the Thompson Trophy Race in both 1938 and in 1939 with his RT-14.

Howard Barlow and John D. Akerman of the University of Minnesota, Turner and the Brown Company completed the new racer in the summer of 1936. Interestingly, Dr. Akerman had also designed a unique tailless aircraft the previous year, though this flying wing flew only once, briefly.

Turner and Brown had differing opinions on the racer's design, however. Brown favored lighter aircraft with small engines, Turner preferred a larger aircraft propelled with as much horsepower as possible. Dissatisfied with the first version of the new aircraft, Turner approached E.M. "Matty" Laird to revamp the aircraft. Laird strengthened the fuselage and fitted a larger wing for better lift and flaps for lower landing speeds.

As completed, the racer, with its powerful 1,000-horsepower Pratt & Whitney Twin Wasp engine, was quickly made ready for the 1937 National Air Races. Turner expected to win; in fact, he had to. With little money, he was counting on his prize-winnings to repay Brown and Laird for the aircraft, known alternately as the LTR, **RT-14**, or a host of other names depending on the sponsor. Unfortunately, while preparing for the Bendix transcontinental event, the racer suffered two fires, one from a leaky fuel line that Turner fortunately blew out with his propeller blast while landing, and the second from a welder's torch while attempting repairs on the fuel tank. He was disqualified. Turner's luck did not improve. While leading the Thompson closed-course race, he mistakenly thought he missed a pylon and circled back. He fell from first to third.

Concerned that the straightforward design of the RT-14 with its fixed landing gear and mixed construction was technologically inferior to the military fighters, such as the P-35 as flown by Jacqueline Cochran, he improved the racer's streamlining and handily won the 1938 Thompson Trophy Race to his great satisfaction. The next year he did the same, becoming the only three-time winner of this prestigious competition. Concerned about his safety at age 43, Turner then promptly retired from racing as had Jimmy Doolittle several years before.

Turner kept his RT-14 on display at his fixed-base operation in Indianapolis and later at his own museum for many years. After his passing in 1970, Roscoe's estate gave his aircraft, his aviation belongings, and his feline friend Gilmore, whom he had had stuffed after the cat's death, to NASM.

Though more mundane than the excitement of air racing and less noticeable than the rapid development of military and commercial aviation, general aviation blossomed during the Golden Age, as the small private airplane filled a variety of important new roles. The Museum is fortunate in having a vast collection of general aircraft of all types, from business aircraft to crop dusters and from personal to homebuilt airplanes.

One early product of the general aviation industry came from France. In 1919, Maurice and Henry Farman, noted especially for their large heavy bombers, built a tiny low-cost biplane for the American market. It was backed by C.T. Ludington, who was soon to finance the creation of National Air Transport and later Eastern Air Lines, and Wallace Kellett, who would turn his

The Farman Sport was an interesting attempt to produce an inexpensive airplane for the American public.

The Bellanca CF was the first cabin monoplane for the civilian market in the United States.

interest toward autogiros. The little 60-horsepower two-seat **Farman Sport** was a delight to fly but it found no market in the postwar recession.

The recession failed to deter other hopefuls. In 1922, Giuseppi Bellanca first flew what is widely believed to be the first cabin monoplane in the United States—the **CF**. Remarkably efficient, the CF featured Bellanca's patented lifting struts that provided lift as well as support for the wing. This strut was later fitted to many aircraft, including some DH-4s operated by the Post Office, with positive results. The CF could seat four passengers inside while the pilot resided in an open cockpit behind the cabin and offset to the right for a better view. With only 90 horsepower, the CF could fly at a remarkable 108 miles per hour and won numerous awards. However, the $5,000 it cost to acquire a CF or a Farman Sport prevented any from being sold as long as surplus Curtiss Jennys were selling for a tiny fraction of that sum.

In 1926, in Havelock, Nebraska, the Arrow Aircraft and Motors Corporation produced its own light plane called the Sport, with 60-, 90-, or 100-horsepower engines. The biplane Sport series featured two-place, side-by-side seating. The wings were cantilevered and of single-piece construction. Conventional "N" struts connecting the wings could be fitted but were

unnecessary. With a price starting at less than $3,000, the Sport was safe to handle with its unusually wide landing gear, had a low 35-miles-per-hour landing speed, and could fly 280 miles while sipping only 4.5 gallons of gasoline each hour. One hundred Sports were sold but Arrow struggled to survive after the onset of the Great Depression in late 1929, eventually closing its doors in 1940.

While many aircraft constructors failed to build a successful light airplane for mass consumption, the Aeronautical Corporation of America—Aeronca—did succeed. Built in Cincinnati, the **Aeronca C-2** made it possible for almost anyone who wanted to fly to do so. Looking like a flying bathtub, the C-2 was a small, inexpensive, high-winged monoplane powered by a tiny 26-horsepower Aeronca engine. Cheap at $1,485 and consuming only four gallons of gasoline per hour, the C-2 reached the mass market of aviation enthusiasts who could not afford the far more expensive and much larger aircraft then being offered to the wealthy. Designed by Jean A. Roche, the French-born senior aeronautical engineer for the Army Air Service, the single-seat C-2 first flew on October 20, 1929, one week before the "Crash" on Wall Street. Despite the subsequent collapse of the economy, Aeronca survived with this excellent little design, carving out a niche market that it dominated for years. Some 164 were built before a factory fire effectively ended its production in the early 1930s. By that time, Aeronca was producing even more popular designs.

The Curtiss-Wright Company of St. Louis, Missouri, also produced a successful light aircraft. The CW-1 Junior resembled a modern-day ultralight and was intended for the pilot on a tight budget. In 1930, two executives of Curtiss-Wright, Ralph Damon (later to become the president of TWA) and Walter Beech (soon to be the founder of his own line of famous aircraft), bought the rights to the design of Orval H. "Bud" Snyder's "Buzzard." Beech, Karl White, and Lloyd Child completely redesigned the diminutive aircraft and installed a three-cylinder 45-horsepower Szekely engine. After test flights in December 1930, the Junior, as the design was now called, was immediately successful despite the Depression. More than 125 were sold within the first six months of production. The price for this open cockpit, two-place, parasol monoplane was almost identical to the Aeronca C-2's at $1,484. The crisis in the economy eventually did catch up, however. By early 1932, when the crisis reached its nadir, sales stopped. With the departure of Beech and other members of the team, Curtiss-Wright retooled the factory to build Condor airliners.

Huff-Daland, a small New York-based company that had built aircraft for the military as well as for civil use, was presented with a unique opportunity that revolutionized agriculture and air travel. In the early 1920s, Dr. B.R. Coad, a government entomologist, conceived the idea that dusting the cotton industry's nemesis, the boll weevil, would be more efficient if it could be done from the air rather than from mule-drawn wagons. Initial trials with Curtiss Jenny biplanes convinced him that researchers were on the right track, but also highlighted the need for an aircraft specially designed for crop dusting. In 1925, the Huff-Daland Manufacturing Company received

The tiny Aeronca C-2 was a light-weight, inexpensive, single-seat private aircraft built during the Great Depression.

the contract to design and produce the Duster biplane based on their military Petrel 5 design.

In 1923, C .E. Woolman, an agricultural engineer and research assistant with the U.S. Department of Agriculture, joined the Huff-Daland Duster Company, located in Macon, Georgia, a subsidiary of Huff-Daland Manufacturing and the first company devoted to dusting operations. Three years later the company relocated to Monroe, Louisiana. Because of the seasonal nature of the crop dusting business, Woolman began to operate in Peru in 1926 during the North American off-season, dusting more than 50,000 acres in Peruvian valleys in 1927. He expanded their business during 1928 by carrying passengers and freight in a Fairchild FC-2 and other aircraft in association with Peruvian Airways Corporation.

Woolman returned to the United States in 1928 and found the firm in financial difficulty. Huff-Daland Airplanes, then part of Keystone Aircraft, wanted to divest itself of Huff-Daland Duster so Woolman took it over, and, with the help of Monroe investors, founded a new company, Delta Air Service, head-quartered in Monroe. Woolman became senior vice president and general manager and planned to carry passengers and freight as well as continuing in the crop dusting business. They inaugurated their passenger and mail service with a route between Dallas, Texas, and Jackson, Mississippi, flying six-passenger Travel Air 6000s. Thus began what eventually was to become Delta Air Lines. Woolman became president of Delta in 1945 and was appointed chairman of the board in 1965. The Huff-Daland Dusters remained in service as dusters until they were replaced by Stearman C3Bs. In 1968,

Delta employees restored and donated the Museum's Duster to honor their founder, C.E. Woolman.

Delta Air Lines owed its creation and success, in fact, to the work of the U.S. Post Office. Between 1918 and 1927, the Post Office blazed the air trails across the country, providing a fast and reliable means of delivering high-priority mail across the country. After the passage of the Air Mail Act of 1925, better known as the Kelly Act, the Post Office turned over the delivery of air mail to private contractors—the airlines, and thus was the U.S. airline industry born. Under the watchful eye of the government, through the Post Office, the industry quickly expanded, soon becoming a vital part of the nation's transportation network and its economy. One of the first companies to respond to the Post Office's request for contractors was Western Air Express.

On April 17, 1926, Western Air Service, Inc., commenced operation on Contract Air Mail Route 4 (CAM-4) between Los Angeles and Salt Lake City via Las Vegas. For service over this route, a distance of about 660 miles, Western selected the **Douglas M-2** aircraft, a mailplane version of the 0-2 observation plane produced by the Douglas Company to replace the U.S. Army DH-4 aircraft.

The Douglas M-2 flew the mail between Salt Lake City and Los Angeles for Western Air Express, now Delta Airlines.

The M-2 performed remarkably well during the early years on the CAM-4 route. Its load-carrying capability, remarkable stability, and rugged construction contributed to a perfect safety record and profitable operation. Government and airline experiences with the Douglas mailplanes and the 0-2 led to modifications of the basic design. Relatively minor changes in cockpit layout, engine accessories, and airframe construction led to the M-3 mailplane, which differed little in physical appearance from the M-2 version. A subsequent addition of five feet to the wingspan resulted in the final

version, the M-4, which realized considerable gain in payload at a negligible loss in performance.

NASM's M-2 is believed to be the last Douglas mailplane in existence. This machine is actually an M-4 model originally purchased by Western from the Post Office Department in June 1927 and registered as NC 1475, serial number 338. The aircraft saw considerable service on Western's mail route until 1930, when it crashed and was sold to Continental Air Map Company of Los Angeles. The airplane had a series of corporate and private owners until it was reacquired by Western Air Lines in April 1940 and subsequently registered with the Federal Aviation Administration (FAA) as M-2 NC15O, Western's first M-2. The first substantial restoration took place in 1946, and for the next 22 years the M-2 made its home in a corner of Western's hangar at Los Angeles International Airport. In 1974, an intensive, large-scale restoration effort commenced under the impetus of retired Western Capt. Ted Homan. The aircraft was restored to flying condition and took to the air on June 2, 1976. After a series of test flights, it was recertified airworthy by the FAA and flown from California to Washington for inclusion in NASM's collection in May 1977.

In the late 1920s, a market developed for larger general-purpose aircraft that could be used for a variety of tasks. Three times more expensive than the Aeronca C-2 and Curtiss-Wright CW-1 Junior, these aircraft were popular with small airlines as well as fixed-base operators. One such design was Hagerstown, Maryland's Kreider-Reisner C-4C. A large open-cockpit biplane with room for three and powered by a 165-horsepower Wright J-6 radial, the C-4C was later known as the KR-34C Challenger after the company was acquired by Fairchild during the merger mania of early 1929. The Interstate Flying Corporation and, later, North Penn Airways flew our

Combining comfort, safety, and utility, the highly successful Beech 18 remained in production from 1937 until 1948.

KR-34C. Some were even outfitted as fighters and flown in combat by the Chinese. After 175 had been purchased, the market for this and other similar aircraft evaporated after the "Crash."

In Troy, Ohio, Elwood Junkin and Clayton Bruckner reorganized the defunct Weaver Aircraft Company (Waco) and in 1925 began production of a popular OX-5-powered biplane for the new Advanced Aircraft Company. Known as the Waco 9, in deference to the memory of the original company, Waco Aircraft, this aircraft became available to barnstormers and other pilots scratching out a living from aviation at a time when the supply of war-surplus Jennys was drying up. Sturdy, maneuverable, and dependable, the Waco 9 sold well and carried mail, dusted crops, and did anything else it was asked to do. Although designed by untrained engineers, the Waco 9 passed the new government certification requirements with flying colors when they were put into effect by the Commerce Department in 1926. When production ended in 1927, 120 Waco 9s had been built.

The Advance Aircraft Company, which now controlled the Waco name, was not finished, however. Because of its well-earned reputation for quality and performance, the Waco name became synonymous with general aviation during the 1930s. Subsequent designs such as the Waco 10 sold well and the company's cabin biplanes achieved great success in the upper end of the market. Wacos such as the UIC in the collection were successful despite the high price tag, often well over $6,000. When the custom UIC first flew in 1931, it introduced aviation to a new market, the businessman. As with Walter Beech the following year, the Advance Aircraft Company/ Waco exploited the need of business to transport its executives rapidly around the country. Though the economy was in tatters, America's top corporations all survived and needed better communications if they were to find a way out of the financial morass. Waco Taperwings and Beech 17s were an answer to the new requirements of business. So successful was the aircraft that Waco could not make them fast enough, selling 73 UICs alone. The UIC featured a comfortable, nicely appointed four-place cabin and was fitted with a 220-horsepower Continental engine—perfect for the travelling businessman who did not wish to depend on airline and train schedules to make his appointments.

Other general aircraft of the Golden Age found success in the burgeoning business market. Starting in 1930 with its popular high-wing Reliant monoplane, Stinson produced a string of highly successful aircraft that catered to the upscale market. By the late 1930s, the graceful SR-10, with its distinctive gull wing and 450-horsepower Pratt & Whitney Wasp Junior, had supplanted the earlier designs. The one in the National Collection, specially modified for a unique service, is currently on loan to the National Postal Museum.

Based on the innovative concept of Dr. Lytle Adams, a Pennsylvania dentist, our SR-10 was modified for the aerial pick-up of mail. This led to the formation of All American Aviation (a predecessor of Allegheny Airlines and today's US Airways), whose task was to deliver the mail by air to

the remote mountain towns and villages along the Appalachian mountains in Pennsylvania and West Virginia as well as in Ohio and Delaware. Six SR-10Cs were first employed in 1939 for this successful service, which entailed flying under contract to the Post Office along a Star route catching suspended mail bags by a hook hanging from the aircraft. Later experiments conducted with our aircraft included the aerial pick-up of humans during World War II. Some success was achieved in September 1943, but the potential was limited.

As previously mentioned, Walter Beech left Curtiss-Wright to form his own company in November 1932. Based in Wichita, Kansas, the Beech Aircraft Company produced its first aircraft in November 1932 in the old Travel Air factory. The Beech 17 was a stunning cabin biplane with staggered wings, one of which is on display in the Golden Age of Flight gallery. The Model 17, also known as the Staggerwing, was an instant success with business-men and also with air racers, who compiled an enviable string of victories that highlighted the aircraft's excellent performance. One variant was capable of flying at 240 miles per hour with just 650 horsepower, a remark-able achievement.

Though Beech was to sell 780 of his Model 17s, he was eager to expand his product line and in 1935 began the design of an all-metal, twin-engined cabin monoplane. The result was the sleek **Beech 18**. Seating seven passengers, the twin-tailed Model 18 was ideally suited for business and small airlines, with which it found remarkable success. It combined the comfort and safety of the modern airliner with the utility and efficiency of a much smaller aircraft. Between 1937 and 1948, when production ended, well over 1,800 Beech 18s were produced for civilian, commercial, and military customers. Many are still flying today.

In Bethpage, New York, the Grumman Corporation created a sleek amphibian to fly the wealthy residents of Long Island quickly and in luxurious comfort to their offices on Wall Street. The **Grumman G-21**, popularly known as the "Goose," proved a successful design and popular with its clients. Soon, the inherent strengths of the G-21 broadened the aircraft's appeal as hundreds were purchased by the Army and Navy as liaison aircraft, while many more entered the commercial market as airliners serving island and coastal routes around the world.

With the nation slowly recovering from the Depression, other manufacturers hoped to return to the idea of an airplane for Everyman. In the late 1920s, brothers C. Gilbert and Gordon Taylor were attempting to sell a little two-seat high-wing monoplane that they had designed and named the "Chummy." When Gordon perished in a crash, Gilbert moved their small company to Bradford, Pennsylvania, where the local Chamber of Commerce was willing to provide sufficient capital to start a new enterprise. With this $500,000 invested, Gordon opened a new Taylor company to manufacture his Chummy. Unfortunately, the Depression ended any hope for prosperity and therefore only five were sold. Undaunted, William Piper, a large stockholder in the company, offered to fund the development of a less expensive version, ultimately known as the E-2. The tiny 20-horsepower engine was not strong enough so the company was forced into bankruptcy in 1931. But Piper was a wealthy oilman. He purchased the assets and kept Taylor on as chief engineer. Fortunately, by this time Continental announced the production of its lightweight, 35-horsepower A-40 engine, powerful enough to get the E-2 in the air and keep it there. With this combination, the E-2 proved an immediate success.

In 1936, the E-2 was redesigned with an improved Continental engine and renamed the Taylor J-2, one of which is in the collection. At the same time, Taylor left to form his own company. Following a fire at the Bradford factory, Piper moved to a former silk factory building in Lock Haven, Pennsylvania, and renamed the company the Piper Aircraft Corporation.

The excellent sales of the J-2 alone—they approached 700 in 1937—prompted William Piper to produce a more powerful version. In 1939, he revealed his classic J-3 Cub, powered by a 40-horsepower engine that was available from several manufacturers. With a sale price of just $1,300, the all-yellow Cub became an instant success. It taught thousands of men and women how to fly, including three-quarters of all the military aviators of the Civilian Pilot Training Program. During the Second World War the Cub, then known as the L-4 "Grasshopper," served gallantly over the battlefield as a liaison aircraft and spotter. By the time production ended in 1947, 14,125 of the legendary J-3s had been built.

The Golden Age of Flight witnessed the immense growth and maturation of aviation into a powerful economic force and an invaluable weapon of war. How critically important this development was to the nation was soon evident when the United States entered World War II.

The Grumman G-21 "Goose" amphibian started as an executive aircraft for the wealthy but grew into a popular all-purpose utility plane and airliner.

The Second World War

The Curtiss P-40 was one of America's frontline fighters at the beginning of the war. This later-model P-40E Warhawk is painted in the colors of the 23rd Fighter Group of the 14th Air Force.

previous spread:
The impeller in the nose of the Messerschmitt Me 163B "Komet" powered the aircraft's electrical system.

Early on the quiet Sunday morning of December 7, 1941, the tranquility of a Honolulu dawn was abruptly shattered when waves of attacking Japanese combat aircraft blasted the U.S. Pacific Fleet at anchor in Pearl Harbor and ravaged other military installations. Caught unaware, the Americans desperately fought back but by the end of the attack most of the U.S. Navy's battleships were sunk or gravely damaged, opening the way for the Japanese conquest of much of Asia and the Pacific. Its complacency shattered, the United States of America was at war.

Present that day in Hawaii were ten Sikorsky JRS-1 amphibians. Those that survived the attack took off in search of the attacking Japanese fleet. Armed with depth charges and bombs, they courageously scoured the seas in the vain hope of striking back. These flying boats were never intended to fight; they were naval versions of the highly successful Sikorsky S-43 commercial airliner that served with distinction on the Latin American routes of Pan American Airways. It was essentially a twin-engined version of the S-42 that surveyed the air routes across the Pacific from Honolulu to Midway, Wake, Guam, and Manila. These U.S. territories were now under siege and unfortunately the S-43 "Baby Clipper" was designed for shorter, less traveled routes. Its success with Pan American had, between 1937 and 1939, led the Navy to purchase 17 S-43s as utility transports. Now with the war in full swing, the JRS was pressed into much more serious duty. They did not last long. By the end of December only one remained. This Pearl Harbor survivor is now in the National Collection.

The Second World War was a crucible in which the tremendous potential of military aviation was fully realized. The strategy and tactics developed in the First World War and honed during the interwar years were put to the test as was the vast array of aircraft designed to fill those roles. Most of the aircraft that fought in the war had been designed in peacetime, but were now undergoing rapid improvements as a result of combat experience. A few were completely new designs. All were thrown into the maelstrom of the world's bloodiest and most violent conflict, where military aviation proved its mettle. During the entire course of World War II, no significant battle was won without control of the skies over the battlefield. Tactical aviation rose to a position of preeminence, becoming a decisive weapon. The strategic bombardment of the enemy's heartland wreaked unimaginable havoc on the military industrial base and the civilian population. Strategic air power, with the ability to transcend the battlefield and carry the war deep behind the lines, expanded the conflict into total war. No longer were there limits on what constituted a legitimate target. The complete destruction of the enemy's will and means to fight was the objective.

America's reaction after the attack on Pearl Harbor was at first stunned disbelief and then grim determination to fight until victorious. Despite the stern resolve that swept the nation, America's fortunes waned throughout early 1942 and reached a nadir with the fall of The Philippines in late spring. Those American forces that survived the last stand at Corregidor either were captured or escaped to fight another day. The stunning setbacks struck hard at American morale as did the revelation that the Japanese were no longer imitators but now innovators in aeronautical design. This came as a great shock. The vaunted Mitsubishi A6M Zero fighter ruled the skies over the Far East with its superior maneuverability, firepower, and climbing ability. When manned by a cadre of well-trained pilots who had gained their experience fighting in China, the Zero was a formidable foe.

But all the news was not bad. In China, a small band of American pilots flying under contract to the Nationalist government was striking back as best they could. The American Volunteer Group (AVG), better known as the "Flying Tigers," first rose up to meet the Japanese two weeks after the attack on Pearl Harbor. They were pilots assembled by the Chinese in 1941 from U.S. Army, Navy, and Marine Corps units and with the blessing of President Roosevelt. Three squadrons of Flying Tigers fought tenaciously against overwhelming odds and for an indifferent Nationalist army to hold the enemy at bay. Their primary weapon was the sturdy and dependable Curtiss P-40. Developed from the pre-war Curtiss P-36, the P-40 flew with a 1,100-horsepower liquid-cooled Allison V-1710 inline engine rather than the P-36's Pratt & Whitney R-1830 Tomahawk, which produced more drag and less power. Though an obsolescent design by the early 1940s, the P-40 gave a good account of itself when flown aggressively.

Under the tutelage of Lt. Gen. Claire Chennault, a pioneer in pursuit tactics, the AVG quickly learned never to turn with a highly maneuverable Japanese fighter but rather to take advantage of the P-40's superior speed, diving ability, and firepower by flying down through enemy formations. This led to the AVG's claim of destroying almost 300 Japanese aircraft during their seven months in action, with a loss of only 11 P-40s in combat. The Flying Tigers flew early B models of the P-40 armed with two .50-caliber machine guns firing through the propeller and four wing-mounted .30-caliber machine guns. By the early summer of 1942, the AVG was equipped with the improved **P-40E Warhawk**, armed with six .50-caliber guns that gave it withering firepower. In July 1942, the AVG was incorporated into the Army Air Forces and eventually fought as the 23rd Fighter Group of the 14th Air Force. Our P-40E is painted to represent a Warhawk of the 23rd Group's 75th Fighter Squadron.

Approximately 14,000 P-40s were built during the war. They served with distinction on every front and with every member of the Allies. Though outclassed by the superior German Messerschmitt Bf-109, the P-40 held its own, particularly in Africa with the Royal Air Force (RAF). The P-40 provided much-needed materiel support for the beleaguered British and enabled the RAF to concentrate its superlative Supermarine Spitfires in home defense during the dark days of the Battle of Britain in 1940.

Two aircraft that met the P-40 in combat over China are in the National Collection: the Mitsubishi G4M "Betty" medium bomber and the Nakajima Ki-43 Hayabusa, better known as the "Oscar" fighter. First flown in October 1939, one year after the P-40, the G4M was an outstanding bomber with an

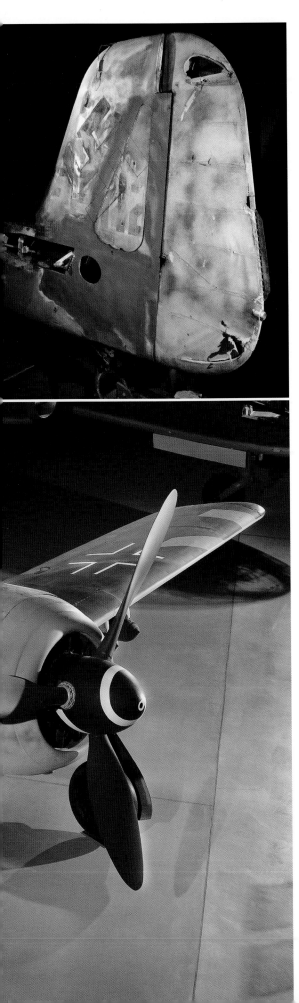

performance suffered, particularly at high altitude. Nevertheless, the P-39 entered service in 1941 and saw combat with units in the southern Pacific and North Africa, where its strength and immense firepower partly offset its inferior performance against later Axis fighters. Consequently, when the Soviet Union applied for Lend Lease assistance, the Army gladly shipped its unpopular P-39s to the Red Air Force. Surprisingly, the Soviet fighter pilots immediately took to the little fighter, which earned an enviable reputation against the best the Luftwaffe could offer. As most of the Red Air Force's combat missions were below 20,000 feet, the P-39 was in its element and, combined with the terrific firepower of its 37 mm cannon and mix of heavy machine guns, proved a highly successful fighter and ground-attack aircraft. In fact, most of the great Soviet aces, particularly Alexander Pokryshkin with 59 confirmed victories, flew P-39s at some stage in their careers. The P-39 was fitted with a modern reflex gunsight and a reliable radio, unlike many of its Soviet contemporaries, and this contributed to its great popularity with Red Air Force. Of the total 9,558 P-39s built, approximately 5,000 were shipped to the Soviet Union. The improved Bell P-63, which featured a new wing, a larger tail, and a more-powerful engine, also saw wide service with the Soviets.

Representative examples of both these aircraft are in the collection. Our P-39 has a special history. Built in 1943, this Airacobra served as an advanced trainer and never went overseas. After the war, Betty Haas (now Betty Haas Pfister) purchased the aircraft and flew it in numerous competitions and air shows, resplendent in its beautiful red and white markings. In 1955, Mrs. Pfister donated her *Galloping Gertie* to the Museum where it will represent both the military and racer history of the P-39.

As the Germans desperately fought on the defensive after their shattering defeat at the Battle of Stalingrad, they found themselves increasingly outnumbered and outclassed by the Red Air Force. In response, the Luftwaffe modified existing aircraft types to combat its numerical inferiority owing to war losses, and to develop desperately needed improved ground-attack aircraft to counter the thousands of excellent T-34 tanks that were pushing the German army westward.

Designed as a replacement for the venerable and obsolete Junkers Ju 87 Stuka dive bomber, the **Focke-Wulf Fw 190F** series entered combat in 1942. Essentially a G-series fighter-bomber with a larger "blown" canopy for better visibility and equipped with almost 800 pounds of increased armor protection, the Museum's Fw 190F-8 possessed impressive firepower. It was armed with two 20 mm cannons and two 7.9 mm machine guns and could carry 4,000 pounds of bombs. It was also capable of launching 24 deadly R4M unguided rockets. The Fw 190 was Germany's only radial-engined fighter, but it proved its worth as a dangerously effective fighter and bomber throughout the war. The BMW 801 14-cylinder radial produced 1,700 reliable horsepower and Luftwaffe pilots found the aircraft to be a more maneuverable and predictable aircraft to fly than its Messerschmitt Bf 109 counterpart.

The legendary de Havilland Mosquito
served with distinction as a recon-
naissance aircraft, a night fighter,
and as a bomber and strike bomber.
Built of plywood, this versatile air-
craft was manufactured in England,
Canada, and Australia. The Museum's
example is a B Mk. 35 bomber
version.

Our Fw 190F-8's restoration was completed in 1983 after years of painstaking work by a team led by the late Mike Lyons, a specialist in German aircraft. His restoration revealed the aircraft's detailed history through the careful removal of paint one layer at a time. The aircraft appears as it did in late-1944 when it flew in Hungary against the Soviets as part of Schlachtgeschwader 2 (Ground Attack Wing 2) at one time under the command of the legendary Stuka pilot and tank destroyer Hans Ulrich Rudel.

The Museum also has two important variants of this aircraft: the Fw 190D-9 and the **Ta 152H**. Developed from the original Fw 190A series, both aircraft were equipped with powerful inline engines instead of the standard BMW radial. The "Dora" series was fitted with a 2,240-horsepower Junkers Jumo 213 engine, normally a bomber engine. This gave the aircraft excellent high-altitude performance, better than the A series and comparable to the super-lative North American P-51 Mustang, against which it was battling for control over the skies of Germany. The Ta 152, so-named after Focke-Wulf's chief designer Kurt Tank, was a further refinement of the "Dora" series. The Ta 152H in the collection featured a pressurized cabin, a more powerful engine, and a longer wing of high aspect ratio. The H was specifically designed to intercept high-altitude Allied heavy bombers and fast reconnaissance aircraft.

The Germans needed all the technological help they could get. By 1943, despite the success of the Focke-Wulf series, the Germans were being pushed back relentlessly, particularly in the East. By early 1944, Great Britain and the United States were striking telling blows at Germany's industrial heart as the Combined Bomber Offensive, begun the previous year, was finally coming into its own with devastating effectiveness, particularly against the Luftwaffe and Axis petroleum production.

Key to this campaign was strategic bombardment, a concept initially developed during the First World War, which called for the destruction from the air of the enemy's factories, communications systems, and transportation network far behind the battlefield. These targets were central to the independent mission of Britain's Royal Air Force (RAF) and the U.S. Army Air Forces during the Second World War. Using large, four-engined heavy bombers, the RAF struck Germany's cities by night in the hope of disrupting entire cities and lowering civilian morale. Meanwhile, the U.S. attempted to strike specific economic and military targets by day, using its famous Norden computational bombsight. Stiff German opposition forced the British to abandon precision daylight bombardment for nighttime area attacks early in the war.

Late in 1943, the Luftwaffe also forced the U.S. Army Air Forces to stand down its bombing campaign until it could re-equip with long-range escort fighters. Nevertheless, by 1944, when the strategic bombing offensive reached its maximum effectiveness, it made a profound contribution to the war effort. Well over one million German soldiers and countless thousands of heavy antitank artillery pieces were pulled from the front lines and brought home to protect German cities from aerial attack. While factories and rail yards were being systematically attacked and Germany's infrastructure

Lockheed P-38 Lightnings were flown by America's top two aces, Thomas McGuire, Jr. and Richard Bong.

disrupted, Allied fighters destroyed most of the Luftwaffe by the late spring of 1944, just in time to clear the skies for D-Day and the invasion of Europe.

Along with the classic four-engined Avro Lancasters, Handley Page Halifaxs, and Short Stirlings, the RAF had a unique aircraft that demonstrated great versatility as a strategic bomber and in many other roles. The **de Havilland DH.98 Mosquito** was an unsolicited design presented by Geoffrey de Havilland to the RAF in 1938 as a high-speed bomber. Built of plywood, the Mosquito astonished the RAF with its top speed approaching 400 miles per hour, as fast as the latest single-engined fighter. Powered by two excellent Rolls Royce Merlin V-12s, the Mosquito gained legendary reputation for versatility when it entered service late in 1941.

Almost untouchable as a reconnaissance aircraft, the DH.98 flew with impunity over enemy-held territory. It served with distinction as a night fighter and was particularly effective as a bomber and strike aircraft. With only a two-man crew it could carry two tons of bombs deep into Germany, and on several occasions gained notoriety for its pinpoint attacks against Gestapo headquarters in Oslo, Norway, and throughout occupied Europe. Equipped as a pathfinder, Mosquitoes flew ahead of British bomber streams, locating and illuminating the target for attack. Because of its speed it was also used as a transport aircraft, carrying ball bearings and high-priority personnel from neutral Sweden across German-controlled airspace with impunity. The Museum's Mosquito is a late-model Mk 35 built in 1945 and given to NASM by the RAF in 1963. Originally intended as a high-speed bomber capable of 422 miles per hour, this version was used as a target tug after the war ended.

For the United States, the Boeing B-17 epitomized its strategic bombing efforts. Flying with the faster but more vulnerable Consolidated B-24 Liberator, American-flown B-17s first struck German targets in the summer of 1942. Based on combat experience learned from the RAF, which also operated the B-17, Boeing modified its original design by adding a tail-gunner position, a dorsal turret, and a ventral ball-turret for much-needed defensive firepower. In addition, a larger fin and rudder from the B-307 transport were installed to improve handling. The B-17E and its improved variant, the B-17F, bore the brunt of unescorted American daylight assaults. Despite striking hard at German aircraft and ball-bearing production at Schweinfurt and Regensberg, the rugged B-17s of the Eighth Air Force were mauled by the Luftwaffe in 1943. By 1944, new G versions with a power-operated chin turret, such as *Shoo Shoo Shoo Baby*, which is destined for NASM's collection, gave additional protection from the effective Luftwaffe frontal attacks.

Despite the increased defensive armament on the B-17G and later versions of the B-24, American daylight raids could not reach deep into Germany until the advent of long-range fighter escort. The loss rates of unprotected bombers reached more than 20 percent on some missions—dangerously higher than the accepted 3 percent ratio. Attempts to provide fighter escort had not proven successful. The unique twin-engined, twin-boom **Lockheed P-38 Lightning** had the range to fly to Berlin but its large size,

The eight-gun P-47D Thunderbolt could carry up to 2,500 lbs. of bombs or a combination of bombs and fuel tanks depending on the mission. The P-47 became the most successful American tactical fighter-bomber of the war.

lack of adequate heating for the pilot, and problems with operating in the cold temperatures of high altitude over Europe greatly restricted its usefulness. The P-38 came into its own later as a ground-attack and reconnaissance aircraft, and even as a successful strategic bomber. In the Pacific, America's top two aces, Richard Bong and Thomas MacGuire, with 40 and 38 victories, respectively, flew P-38s with great success. NASM's late-model P-38J remained in the United States during the war and was transferred from the Army Air Forces to the Museum in 1946.

Unlike in the Pacific, the Lightning was not suited for the air war in northern Europe. The aircraft that rescued the American strategic bombardment campaign was the legendary North American P-51, one of which is in the World War II gallery in the Museum. When it entered service with the Eighth Air Force in December 1943 and flew its first deep escort missions over Germany a few months later, even Hermann Goering knew the tide had turned irrevocably in the Allies' favor. Sleek, fast, and maneuverable, the Mustang could carry the fight over the heart of Germany and defeat the best the Luftwaffe had to offer. Interestingly, the Mustang's rotund prede-cessor, the **Republic P-47D Thunderbolt**, actually destroyed far more German aircraft than the Mustang, in fact more than any other fighter, and was responsible for the defeat of the Luftwaffe in the spring of 1944.

Developed from the Seversky P-35, the Republic P-47 first flew in May 1941. Big and powerful with a supercharged, 2,000-horsepower Pratt & Whitney R-2800 radial and armed with eight .50-caliber heavy machine guns, the "Jug," as the P-47 was popularly known, earned a well-deserved reputation for ruggedness. It was capable of out-diving any opponent and despite its large gross weight of almost 15,000 pounds it could fly well over 400 miles per hour and had a remarkably high roll rate. In the hands of well-trained pilots, such as those of the Eighth Air Force's 56th Fighter Group who flew the Thunderbolt throughout the war, the P-47 was a formidable foe. Most important, improvements in the aircraft by early 1944, particularly the installation of larger external fuel tanks, allowed the P-47 to fly as escorts on all but the longest missions. The Thunderbolt is credited with the destruction of 7,067 enemy aircraft on every front of the war. NASM owns a P-47D-30-RA that remained in the United States after it was delivered to the Army in October 1944. It is typical of those Thunderbolts that so effectively ravaged the Luftwaffe.

In 1944, the Army Air Force discovered a new and unexpected role for the P-47. Although it was designed as a high-altitude interceptor and escort fighter, the immense strength and durability of the P-47 made it an ideal ground-attack aircraft. Under orders to fly "on the deck" when returning from escort missions, P-47s excelled in destroying targets of opportunity. The Thunderbolt's withering firepower and ability to carry large bombloads and, later, rockets, made it a formidable weapon against the German army. The Ninth Air Force, at the time the largest single air force in the world, specialized in tactical aviation and used the P-47 with great effectiveness in supporting the Allied advance through France after D-Day. Flying close-support missions, P-47s and other aircraft, such as the Douglas A-26 Invader, one of which is also in the collection, were instrumental in destroying the German army on the Western Front.

As excellent as these aircraft were, they would have been useless had they not been flown by highly skilled and highly trained pilots and crew. The United States built more than 300,000 aircraft during the course of the war and trained more than enough crew to fly them. This was a remarkable achievement considering that most of the pilots and crew had never flown before they entered the military. To train these pilots, the military purchased a variety of primary, basic, and advanced trainers. The Museum possesses a North American AT-6 as well as its naval equivalent, the SNJ.

Of particular interest is the recently acquired **Ryan PT-22**. Unique as the only monoplane primary trainer purchased against the backdrop of thousands of Stearman PT-13s and PT-17s, the Ryan PT-22 Recruit was developed from the earlier PT-16, which was a military version of the popular Ryan S-T series of light aircraft. The PT-22 was the first low-wing monoplane used for primary pilot training and made for a smoother transition to more demanding low-wing fighters during World War II. The Army Air Forces accepted 1,023 PT-22s. Ryan also built additional aircraft for the U.S. Navy, and as part of Lend Lease contracts with China and other Allies. Our Recruit was originally the third of 25 built under contract as a floatplane trainer for use in the Netherlands East Indies (NEI) but the sale fell through after the NEI surrendered to Japanese forces in May 1942. It then was used as an Army trainer until it was declared surplus late in the war. Since 1944, this Recruit had nearly two dozen owners before it was donated to the Museum by John Damgard in 2006.

The Ryan PT-22 was the only monoplane primary trainer used by the U.S. military during the war. It featured two cockpits and a beautiful all-aluminum fuselage.

The U.S. Navy faced a similar problem in training the thousands of pilots it needed. Together with the Stearman N2S, the Naval Aircraft Factory in Philadelphia was given the task of manufacturing a new Navy primary trainer in the mid-1930s. Following successful tests, this little biplane trainer was built in both land and seaplane versions. The Navy initially ordered 179 N3N-1 models, and the factory began producing more than 800 N3N-3 models in 1938. U.S. Navy primary flight training schools used N3Ns extensively throughout World War II. A few of the seaplane version were retained for primary training at the U.S. Naval Academy. In 1961, they became the last biplanes retired from U.S. military service. The Museum's N3N-3 was transferred in 1946 from Cherry Point, North Carolina, to Annapolis, Maryland, where it served as a seaplane trainer. It was restored and displayed at the Naval Academy Museum before being transferred to NASM.

Toward the end of the war, hopelessly outnumbered in manpower and supplies and severely pressured on two fronts, the Germans desperately hoped for a technological miracle to salvage their chance for victory. While most of Hitler's schemes were too far-fetched for serious consideration, some of the last-ditch efforts did hold great potential. Some were innovative developments of conventional technology and some explored the limits of aeronautics and propulsion.

To combat the steady stream of Allied bombers that were incessantly pummeling the Third Reich, the Luftwaffe fielded many aircraft later in the war, two of which reside in NASM. A conventional design, the Messerschmitt Me 410 Hornisse (Hornet) had a checkered history as a twin-engined heavy fighter after it entered service in mid-1943. Developed from the unstable and unsuccessful Me 210, the Me 410 was fast and capable of carrying up to four 20 mm cannons, or two 30 mm, or even one powerful 50 mm cannon, any of which could destroy any heavy bomber with just a few bursts. Unfortunately, the day versions of this aircraft were not very maneuverable and were easy prey for American escort fighters. Consequently, the Me 410 was modified to carry airborne radar for night fighting, a role that better suited the large aircraft.

A better, specially-built night fighter, the **Heinkel He 219 "Uhu" (Owl)** entered service in June 1943 and quickly gained a reputation as the Luftwaffe's best weapon against the RAF's Bomber Command. Fast and maneuverable, even for a large twin-engined aircraft, the He 219 easily exceeded 400 miles per hour at altitude and was one of only a handful of aircraft that could intercept the superlative de Havilland Mosquitoes. It was the first operational aircraft fitted with an ejection seat and the first German aircraft with a tricycle landing gear. Few heavy bombers could withstand the withering fire from the Uhu's four 20 mm cannon, and were it not for production delays far more He 219s would have been built—with a correspondingly increased British casualty list.

Designed specifically as a night fighter, the Heinkel He 219 "Uhu" (Owl) wreaked havoc against the Royal Air Force's Bomber Command in the last year of the war.

Based on the excellent airframe of the twin-engined Ju 88 medium bomber, the Junkers Ju 388 permitted the Luftwaffe to develop—at maximum speed and minimum risk—a high-performance, four-seat reconnaissance aircraft and heavy interceptor. Powered by two supercharged, 1,800-horsepower BMW 801 radial engines, the Ju 388 answered the Luftwaffe's request, entering service in the late summer of 1944. Although a derivative design, the pressurized aircraft was an excellent high-altitude photo-reconnaissance platform and proved to be a very capable night fighter, of similar performance to the superlative He 219.

The Dornier Do-335A-1 "Pfeil" (Arrow) tandem-engine day-bomber and fighter-bomber. The fighter-bomber version was designed to carry 1,100 lbs. of bombs. The aircraft was not encountered operationally by Allied air forces.

More exotic, though still powered by conventional piston engines, the **Dornier Do 335A-1 "Pfeil" (Arrow)**, of which NASM owns the last survivor, was a unique twin-engined fighter. With Daimler DB 603 engines of 1,750 horsepower in the nose and in the tail, the Pfeil was intended to fly as a heavy day and night fighter. The unique tandem-engined configuration gave the aircraft exceptional performance, low drag, and a top speed approaching 500 miles per hour. Of all-metal construction and fitted with a tricycle land-ing gear, the low-winged Pfeil possessed a large cruciform-shaped tail. It was also equipped with an early ejection seat that, when activated, ignited explosive bolts in the canopy, rear propeller, and upper fin, clearing the way for the pilot in an emergency. Unfortunately for the Luftwaffe, the first version of the seat was overpowered, with dire results for the pilot.

Although it first flew in September 1943, the massive 21,000-pound Do 335 never became operational because of conflicting and confusing decisions by the Air Ministry. It nevertheless remains an excellent example of the zenith of piston-engined technology. The Museum was fortunate when the Deutsche Museum of Munich offered to restore the Dornier, which they did during the 1970s and 1980s. The aircraft remained on loan in Germany after its beautiful restoration until it was eventually returned to NASM in 1990.

Sophisticated though the Do 335 was, it still had an evolutionary design that used conventional power plant and aeronautical technology. What the German aeronautical industry gained a deserved reputation for was the development of new engine technologies, particularly the jet and the rocket. On August 27, 1939, the Heinkel He 178, piloted by Erich Warsitz, became the first jet aircraft ever to fly. Its radical new engine was created by Hans Pabst von Ohain, a brilliant 26-year-old engineer who developed the turbojet engine concurrently with, but independent of, Frank Whittle in Britain. Because of the immediate pressures of the war, the German Air Ministry initially showed only a lukewarm interest in this private project from Heinkel; the Air Ministry was already funding gas turbine research by BMW and Junkers.

The Air War
in the Pacific

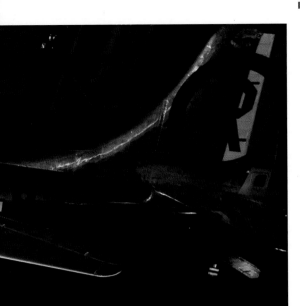

The Grumman F6F Hellcat was the most effective Allied aircraft against the Japanese during World War II.

previous spread:
The Lockheed P-38 Lightning was an effective fighter and bomber that possessed high speed and great range.

T he Allies had concentrated their efforts to defeat Germany, which was seen as a more immediate threat once the Pacific war stabilized. In 1943, with its immense production capability and prodigious organizational prowess bearing fruit, the United States took the offensive against Japan. Armed now with a seemingly endless supply of superb carriers and aircraft, the U.S. Navy struck back hard, and with the Marine Corps and the Army fighting heroic struggles on numerous strategic islands across the Pacific the tide turned in America's favor.

Leading the war was a new generation of naval aircraft faster and more powerful than anything before. In August 1943, a new fighter from the famous Grumman works entered combat for the first time against Japanese positions in the Marcus Islands. Designed in 1941 and based on combat experience, the **F6F Hellcat** soon proved itself to be the most capable naval fighter of the war. With the largest wing area of any single-engined World War II fighter, and powered by the strong and reliable, 2,000- horse-power, Pratt & Whitney R-2800 Double Wasp radial, the Hellcat was fast and very maneuverable. A stable gun platform, it was also a very easy aircraft to fly for the thousands of newly trained naval aviators. It soon demonstrated its preeminence, eventually downing almost 5,000 Japanese aircraft—75 percent of the total destroyed—while losing little more than 250 Hellcats in combat. The portly F6F Hellcat, as rugged as the F4F Wildcat that it replaced, was armed with six .50-caliber M-2 Browning machine guns that were lethal to any enemy aircraft.

The Hellcat's greatest day came on June 19, 1944. During operations in support of the Marine Corps' and Army's attacks on Saipan, Tinian, and Guam in the Marianas Islands chain, U.S. Navy Hellcats fought off successive waves of Japanese aircraft from the island and particularly from the approaching Japanese fleet. After the day had ended, 270 enemy aircraft had been destroyed at a loss of only 26 Hellcats. This overwhelming American victory was soon dubbed "The Marianas Turkey Shoot," as the poorly trained and outnumbered Japanese pilots and their aircraft were no match for the superior Hellcat.

Our Hellcat is an early F6F-3 model that was built in February 1944. Assigned training duties in Hawaii, this Hellcat eventually was converted into a radio-controlled target drone. On one particular postwar mission in July 1946, it was flown through the atomic cloud over the Bikini Atoll after the detonation of a nuclear weapon during Operation Crossroads. Coincidentally, it was during these tests that the captured German cruiser *Prinz Eugen* was sunk. This was the same ship from which our Arado Ar 196 was salvaged. Our Hellcat was thoroughly cleaned and decontaminated; it is not radioactive.

During that two-day naval battle in 1944, later named the Battle of the Philippine Sea, the U.S. Navy was able to deploy a huge fleet of ships and aircraft. Task Force 58 included 15 aircraft carriers, 7 battleships, and more than 900 aircraft, and it destroyed two carriers and most of what was left of Japan's naval air power. A total of 476 Japanese land- and sea-based aircraft and 450 pilots were lost. American naval air power was relentlessly pounding the Imperial Navy. It was only a matter of time before Japan would fall.

Instrumental in this American air offensive was a new and even more effective dive-bomber, the **Curtiss SB2C Helldiver,** which was replacing the venerable Dauntless in the U.S. Fleet. By October 1944, culminating with the Battle of Leyte Gulf during the U.S. liberation of The Philippines, the Helldiver had overcome most of its significant teething pains and had become the principal dive-bomber in the inventory.

Capable of carrying up to 2,000 pounds of bombs—almost twice as much as the SBD—the Helldiver had first flown in December 1940 but did not enter service until three years later. The design had been plagued by a host of problems, particularly the short-coupled fuselage, the length of which was dictated by the size of the elevators on board aircraft carriers. This created a myriad of aerodynamic difficulties, including wing failures and instability. Known as the "Beast" when it entered service, the Helldiver nevertheless demonstrated its capabilities as a weapon throughout the critical battles of the last two years of the Pacific war. Helldivers participated in the destruction of the great battleships *Musashi* and *Yamato* and obliterated what was left of Japan's carriers.

The Curtiss SB2C Helldiver replaced the venerable Douglas Dauntless as the U.S. Navy's primary dive-bomber in 1944. Our Helldiver will be the first aircraft restored at the Mary Baker Engen Restoration Hangar.

During the lengthy campaign to free The Philippines, Helldivers and other American aircraft rampaged throughout that island nation in search of targets. During the Battle of Leyte Gulf, a young Lt.j.g. Donald D. Engen flew from the USS *Lexington* as a member of bombing squadron VB-19 in pursuit of Japanese Admiral Ozawa's fleet. Although most of the Japanese navy's aircraft had been destroyed earlier at the Battle of the Philippine Sea, Ozawa's three carriers and two battleships still remained a dangerous threat. On October 25, 1944, Engen and his gunner took off with the other 11 SB2C-3s of his squadron, along with 12 Grumman TBFs of VT-19 and 36 escorting Grumman F6Fs. They were assigned to strike the largest enemy carrier, the *Zuikaku*, a veteran of the attack on Pearl Harbor. Following the other members of his squadron, Engen dove on the carrier through intense antiaircraft fire placing his 1,000-pound bomb right on the target. Pulling out of his dive only 25 feet above the water, Engen flew his rugged Helldiver under the bow of the battleship *Hyuga* and noticed that its officers were standing on its bridge in their dress white uniforms. For his participation in the sinking of this fleet carrier, Engen was awarded the Navy's highest decoration, the Navy Cross. Two weeks later, his squadron struck another target in Manila Bay. Nosing his large bomber over into a steep 60-degree dive, Lieutenant Engen pointed his Helldiver directly at the heavy cruiser *Nachi*, sinking it as his bomb and those of VB-19 rained down on the doomed warship.

After the war, Donald Engen went on to a distinguished career, retiring as a vice admiral before leaving the service and entering civilian life as Federal Aviation Administration Administrator, business executive, aircraft safety expert, and eventually director of the National Air and Space Museum. He passed away in a sailplane accident in July 1999.

Engen and other Navy dive-bomber pilots were escorted to their targets not only by the ubiquitous Grumman F6F Hellcat but also by what is widely considered one of the finest all-around fighters of World War II, the **Vought F4U Corsair.** America's first production aircraft to exceed 400 miles per hour, the Rex Beisel-designed Corsair was powered by a 2,000-horse-power Pratt & Whitney R-2800 radial as was the Hellcat. The 425-mile-per-hour Corsair was some 50 miles per hour faster than the Hellcat in part because its sleeker design and unique gull-wing allowed the installation of a larger-diameter propeller for maximum efficiency.

Were it not for the aft placement of the cockpit, which hampered visibility on takeoff, the Corsair would have entered carrier service with the U.S. Navy in 1943. The first F4Us were issued to land-based squadrons instead, particularly those of the U. S. Marine Corps, which used the Corsair's superior speed and firepower to devastating advantage, achieving an 11:1 victory ratio over the enemy. Maj. Gregory "Pappy" Boyington of Marine fighter squadron VMF-214, the famous "Black Sheep" squadron, became the Marines' leading ace with 28 victories. Incidentally, during the restoration of NASM's F4U-1D, Major Boyington visited the shops and signed his name inside the starboard main gear well.

The Vought F4U Corsair was a highly successful land- and carrier-based fighter and fighter-bomber that performed exemplary service in World War II and Korea. The F4U-1D seen here appears to bank away from the NAF N3N trainer above the Lockheed SR-71 in the Boeing Aviation Hangar.

The Corsair excelled as a fighter but also proved itself to be an exceptional bomber, capable of carrying two 1,000-pound bombs and other heavy ordnance. Charles Lindbergh, the famous civilian aviator, who at that time was a representative for United Aircraft, demonstrated to astonished Marine pilots that the Corsair could carry even more. Lindbergh successfully dropped 4,000 pounds of bombs in a single attack on the island of Wotje, demonstrating the immense strength of this remarkable aircraft.

Following the lead of the Royal Navy's Fleet Air Arm, which successfully flew their Corsairs from carriers in 1943, including the Corsair's first sortie—against the German battleship *Tirpitz*, the U.S. Navy relented and began carrier operation in 1944. It was a wise decision. By the end of the war, more than 12,000 Corsairs had been built, and the aircraft had earned a well-deserved reputation as a superlative multipurpose fighter that played a significant role in the Allied victory.

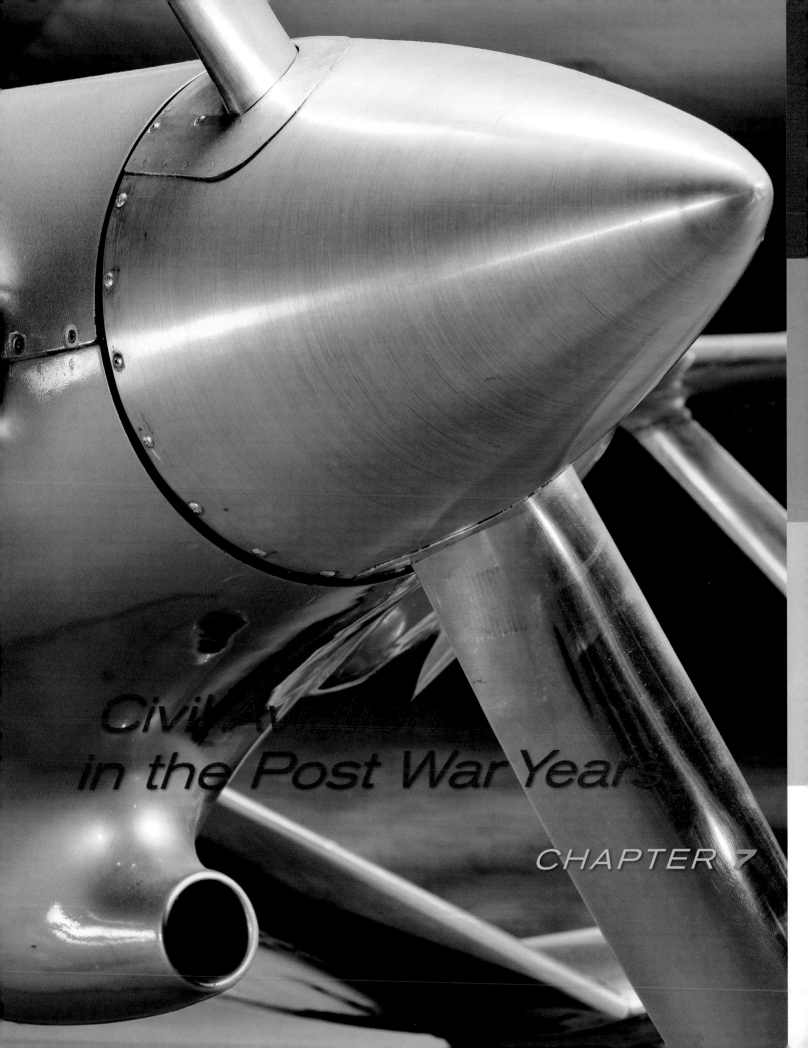

Civil Aviation
in the Post War Years

CHAPTER 7

Flush in the immediate glow of victory, America quickly retooled its economy to take advantage of its newly found economic power. Within a year of the end of the Second World War, the nation's economy began to expand rapidly, as years of pent-up demand for consumer goods and enforced savings were unleashed in a consumer-driven economic boom. Returning veterans were granted unprecedented privileges through the GI Bill of Rights that gave millions of citizens access to low-cost home loans and unheard-of educational opportunities. This prompted a massive building boom especially in newly-created suburbia and flooded the nation's colleges and universities with new students. America entered a period of prosperity that would last for more than two decades.

As consumer spending rose dramatically, many leaders in the aviation industry sought to capitalize on this emerging market. With the cancellation of billions of dollars worth of war contracts, the aircraft manufacturers were banking on the expansion of the general aviation market to offset these losses. Thousands of veterans who had learned to fly during the war were expected to continue in aviation. Business leaders hoped that these former GIs would spend their money on new private aircraft that were beginning to arrive on the market. These sanguine hopes spurred a large expansion of the civil aviation industry, but the industry's hopes were soon dashed. Most former military pilots just wanted to get back to their peacetime careers and build a home for their new families. An aircraft for their personal use was out of the question.

After this early struggle, however, the aircraft manufacturers were able to carve out a comfortable market for their general aircraft. By the 1950s, general aviation was expanding as the nation's prosperity during the Eisenhower years produced an unprecedented amount of discretionary income. Some of this was spent on aircraft, and not just automobiles and televisions. New uses for these aircraft also increased their utility. Originally built for personal transportation, general aircraft were found to be ideally suited for a variety of tasks from aerial taxi service to crop dusting to survey work. Air racing returned—though on a smaller scale, sailplanes increased in popularity, and businesses took a renewed interest in finding faster ways to transport executives quickly around the country.

By the 1970s, with the cost of personal aircraft skyrocketing amidst inflation and lawsuits, the low-cost homebuilt aircraft industry found a ready market, and NASM's National Aeronautical Collection preserves many examples of these important aircraft.

Following the conclusion of the Second World War, manufacturers such as Piper, Beechcraft, and Cessna, among others, returned to the production of civil aircraft. The trend toward more versatile, all-metal design begun in the late 1930s continued with a host of new and improved aircraft that quickly reached the market. Concurrent with the rising interest in general aviation, many sought to publicize the capabilities of these new aircraft through a series of record-setting flights. Others just came and went with the changing economy.

Betty Skelton's Pitts Special S-1C **Little Stinker** *now greets visitors as they enter the Steven F. Udvar-Hazy Center.*

previous spread:
The Mahoney Sorceress is the most successful biplane in air racing history.

In 1930, in Riverdale, Maryland, a quiet suburb of Washington, D.C., the small Engineering and Research Corporation (ERCO) was created by Henry Berliner to manufacture precision tooling machines for aircraft companies. ERCO made controllable-pitch propellers, aluminum sheet formers, and propeller profilers. In fact, all U.S. propellers built during World War II were ground on ERCO profilers.

In 1937, Berliner decided to enter the light aircraft business and hired noted National Advisory Committee for Aeronautics (NACA) engineer Fred Weick as chief designer. Weick had already built his W-1 and W-1A high-wing pusher for a safe airplane contest sponsored by the U.S. Commerce Department. Berliner decided that a more conventional aluminum low-wing tractor design would be more marketable, so the two combined their considerable expertise and produced a unique two-seat, low-wing monoplane they named the **Ercoupe.** The first version was the ERCO 310 and featured a twin tail and a tricycle landing gear, a first for a light airplane. This evolved into the production 415-C that was powered by a 65-horsepower Continental A-65 "flat four" engine.

What made the Ercoupe unique was its original control system, which eliminated the need for rudder pedals. The rudders were linked directly to the full-length ailerons, thus ensuring well-coordinated turns. The careful design of the wings and the limited travel of the elevators ensured that the Ercoupe was immune to stalls and spins. Training time for a new pilot was half that for a conventional aircraft, making the Ercoupe an immediate success with hundreds of casual flyers. However, as it was fitted with a two-control system, rather than three, which featured a conventional control wheel but no rudder pedals, Ercoupe pilots received a special license from the Civilian Aviation Authority restricting them to this single aircraft type. Eventually ERCO relented and offered Ercoupes with a traditional three-control system as well. Undaunted, Erco produced 112 Ercoupes before World War II interrupted production and the company shifted its efforts to the war industry.

After the war, ERCO resumed production of the Ercoupe to fill the thousands of orders that were expected to come from the returning servicemen. The Ercoupe 415-D had a more powerful 75-horsepower Continental engine. Other versions soon followed. For his pioneering work in building a safe, popular light aircraft, Fred Weick received the Fawcett Aviation Award. The Ercoupe developed a loyal following but even its devotion was not enough to overcome slow sales. By the outbreak of the Korean War, ERCO had withdrawn from the light aircraft market and returned to its more lucrative tooling business. The rights to the Ercoupe were sold to Forney Manufacturing of Fort Collins, Colorado, which built the aircraft until 1959. After a succession of owners, the Mooney Aircraft Corporation, the last builder of Fred Weick's innovative design, finally ended production of this clever little aircraft in 1970.

Built in Riverdale, Maryland, by the Engineering and Research Corporation, or ERCO, this Ercoupe is seen flying over Washington, DC. An earlier version of this aircraft was the first light airplane to be fitted with tricycle landing gear.

NASM is fortunate in owning the first production Ercoupe 415-C ever built. It was completed in October 1939 and received its airworthiness

Piloting her Cessna 180 Spirit of Columbus in 1964, Geraldine Mock became the first woman to fly around the world.

realized the need for a smaller, less-expensive aircraft to satisfy the demand in the expected coming postwar market boom. The result was the metal, two-seat Cessna 140 featuring a strut-braced wing that was partially covered in fabric and a small, air-cooled, four-cylinder, 75-horsepower Continental engine. It was also fitted with the spring-steel landing gear pioneered by Steve Wittman, which was to become a Cessna trademark. Sales of this attractive light aircraft were brisk. A stripped version known as the 120 also met with success, as did later improvements.

Cessna realized, however, that the market required a four-seat aircraft to satisfy the demand from businessmen and private pilots who wished to fly their friends and families. In 1947, Cessna produced its Model 170 based on the 140 but with seating for four and a more powerful 145-horsepower, six-cylinder Continental C-145. By 1949, the Model 170 evolved into an all-metal design and eventually metamorphosed into the outstanding Cessna 172 after its conversion to tricycle landing gear. The 172 has been one of the most widely produced aircraft of any kind in history with almost 35,000 built.

Despite the success of the 170 and 172, Cessna realized the need for an even larger aircraft. As a result Cessna enlarged the existing design of the Model 170 in 1953 to produce the Model 180. With a flat-six, air-cooled, 225-horsepower Continental O-470-A engine, the 180 sold well and became especially popular with bush pilots for its sturdiness and ability to carry heavy loads in difficult terrain. A later tricycle-geared version became even more popular as the Cessna 182.

The National Air and Space Museum owns a special **Cessna 180**. It was owned by Mrs. Geraldine Mock, a private pilot from Columbus, Ohio, who in 1964 officially became the first woman to fly an aircraft around the world.

Her Model 180, the *Spirit of Columbus*, was the 238th Model 180 built and was modified with additional fuel tanks mounted inside the cabin. On March 19, 1964, "Jerrie" Mock took off from Columbus and headed east. Almost one month later, on April 17th, she returned to Columbus, having flown 23,103 miles in 29 days, 11 hours, and 59 minutes while setting a new speed record in her aircraft's weight class. For this flight, President Lyndon Johnson awarded her the Federal Aviation Administration (FAA) Exceptional Service Decoration.

To honor her flight, Cessna gave Mock a new aircraft in exchange for her *Spirit of Columbus*. For years this special 180 was displayed at the factory until in 1975 Cessna refurbished the aircraft and presented it to the Museum.

One of the most interesting aircraft preserved at NASM is the unique Windecker Eagle. In the late 1950s, Dr. Leo Windecker, a dentist from Texas, together with his wife, first envisioned a light aircraft made from unconventional materials. In cooperation with the Dow Chemical Company, Dr. Windecker developed a fiberglass-reinforced plastic he dubbed "Fibaloy." Extremely resistant to heat and cold, this material could be cast easily yet was very strong, could be glued, and was readily machined. After conducting extensive tests on a Luscombe Monocoupe that he modified with a set of Fibaloy wings, in 1967 Dr. Windecker formed Windecker Research, Inc., in Midland, Texas, to build a sleek, low-winged, four-seat aircraft.

The company's first product was the X-7, which featured a fixed landing gear and was used as a flying test bed to demonstrate the suitability of the new material. After successful flight tests, the X-7 was stripped of its engine and accessories and parked outside in the elements to serve as a passive laboratory to determine the long-term characteristics of Fibaloy when exposed to the elements.

Having gathered considerable information from these experiments, Dr. Windecker forged ahead with his dream to build a practical aircraft. In 1969, the first prototype of his Eagle made its first flight, propelled by a 285-horsepower Continental IO-520 engine. FAA concerns about its strength and durability forced Windecker to build the aircraft to requirements 20 percent stronger than those required for conventional all-metal designs. The FAA had little to fear. Despite the loss of the prototype in an accident, the Fibaloy was lighter and stronger than aluminum and actually increased in strength as it aged. The Eagle received its type certificate in December 1970.

The Eagle was a technological success. Fast and efficient, the aircraft should have been a winner. Unfortunately, the light aircraft market in 1970 was severely hurt by a recession, escalating costs, and spiraling litigation expenses that cut sales. Soon after the first production Eagle was delivered, Windecker's financial backers withdrew their support and the enterprise closed its doors after only six Eagles were sold.

That should have been the end of the story, but in 1973 Gerald P. Dietrick became intrigued with the Eagle after buying one second-hand. He contacted Dr. Windecker to see if the company could be resuscitated. Although

This Piper PA-18 Super Cub was flown by the Atomic Energy Commission to search for uranium.

rebuffed by potential backers, Dietrick purchased the physical assets of Windecker Industries as well as the type certificate and in 1979 formed the Composite Aircraft Corporation. In order to demonstrate the capabilities of the Eagle, Dietrick flew his Eagle to the Paris Air Show that year in a record 21 hours and 58 minutes, averaging 193.35 miles per hour and breaking a 12-year-old record previously held by a Beech Bonanza for aircraft in its class. Unfortunately, even this demonstration was insufficient to win investors and, in 1985, the Dow Chemical Company donated its Windecker Eagle to the Museum.

The idiosyncratic Willard Custer of Hagerstown, Maryland, produced a more extreme general-aircraft design. In the 1950s, Custer became enamored with the idea of designing and building an aircraft fitted with a special, high-lift wing. Inspired by the dramatic lifting ability of the wind when he witnessed the removal of a barn's roof during a storm, he began to experiment with a series of designs until settling on a semi-circular channel shape. By directing the propeller wash over these unique surfaces, Custer's "Channel" wings were able to generate considerable lift. The CCW-1 was capable of short takeoff and landing performance with excellent low-speed handling even with one of its two 75-horsepower Lycoming engines out.

While the design held promise, the weight of the structure and the drag, together with Custer's difficulties in finding suitable financial backing, prevented him from achieving success. In 1962, Custer gave his first

The Waterman Aerobile is technically a flying motorcycle with its three-wheeled landing gear and detachable wings.

Channelwing, the CCW-1, to the Museum while he was working on subsequent aircraft. His attempts to sue Fairchild Republic over the design of the U.S. Air Force A-10 Thunderbolt II attack aircraft for patent infringement failed. He claimed that the A-10's rear-mounted turbine engines drew air across the wing to provide lift in the same manner as his Channelwing. The courts thought differently.

NASM also preserves other light aircraft such as a **Piper PA-18 Super Cub**, used by the Atomic Energy Commission to search for uranium, and an example of the ubiquitous Cessna 152 that has achieved widespread success as a trainer and commuter aircraft since the late 1950s.

General aviation entails more than just conventional light airplanes. That diversity is reflected in other aircraft in the NASM collection. For example, one of man's perpetual yet chimerical dreams is that of a flying automobile. Countless inventors, serious engineers, and dilettantes alike, have attempted to combine automobile and aircraft technologies with varying

degrees of success. Several of these often whimsical designs reside in the study collection.

Built during the Great Depression by early aviator Waldo Waterman, the **Waterman Aerobile** featured a tailless fuselage with easily detachable wings. The Aerobile was registered as a motorcycle because of its tricycle landing gear. A single 100-horsepower Studebaker automobile engine mounted in the rear drove a pusher propeller to provide the power on the first variant. Because a plan to sell the Aerobile through Studebaker dealerships never materialized, only six of these hybrid flying cars were built.

The Bureau of Air Commerce took an even more novel approach in 1935 when it awarded an experimental contract for the construction of a road-worthy autogiro. As a result, the Autogiro Company of America built six AC-35s based on the Pitcairn PA-22, the first of which flew on March 26, 1936. Of mixed wood-and-metal construction and covered in fabric, the AC-35 incorporated folding wings for road use and garage storage and was driven by a 90-horsepower Pobjoy engine that gave the small craft a top road speed of 25 miles per hour. Unfortunately, no market was found for this vehicle so production was limited to six examples.

After World War II, interest in flying cars was renewed in many guises. One of the most intriguing was the design of Robert Fulton, a businessman who grew tired of not having an automobile at the local airports he frequented. He flew the first prototype of his **Fulton Airphibian** in 1945. The aircraft consisted of a semi-monocoque forward fuselage that doubled as an automobile. When operated as an aircraft, the cabin section was attached to a steel tube, fabric-covered rear fuselage and tail section. The Fulton also used a cantilevered metal and fabric wing mounted above the forward fuselage, as in the later FA-3-101 that is in the NASM collection. A single Franklin 6A4 engine produced 165 horsepower. This gave the Airphibian a top air speed of 110 miles per hour and a road speed of 55 miles per hour. In 1950, this vehicle received the first approved type certificate for a flying car. Eight were built but high manufacturing costs prevented the acquisition of sufficient capital to begin mass production. Consequently, the Fulton Airphibian never was widely produced.

Amphibians, though equally interesting, are perhaps a more practical than flying cars. These aircraft combine the advantages of a flying boat with the convenience of retractable wheels for conventional landings and take-offs. Typical of this particular aircraft was the Republic RC-3 Seabee that for several decades found a small niche in the general-aviation market. Designed for safety with its 215-horsepower Franklin 6A8 engine and Hartzell controllable-pitch propeller installed behind the enclosed four-seat cabin, the all-metal Seabee was first flown in 1945. It was produced for only two years but was so popular that more than 1,000 examples were sold before the postwar recession curtailed production. Though underpowered, the Seabee gained a loyal following as its unusual design permitted considerable flexibility in operation.

In 1950, the stunning red Fulton Airphibian received the first approved-type certificate for a flying car.

This Bücker Bü 133 Jungmeister aerobatics biplane was flown by champion pilot Beverly "Bevo" Howard.

Ever since the beginning of aviation, air shows have thrilled crowds around the world. Soon after the Wright brothers took to the air, they and countless imitators sought to demonstrate the superiority of their products through aerial demonstrations. First by racing and later by daring aerial maneuvers, they captivated the attention of aviation enthusiasts. In addition to converted military fighters, such as Al Williams's *Gulfhawk*s, specially-built aircraft were eventually produced to sustain the high stresses of aerobatics. By the 1930s, and especially after the Second World War, several purpose-built civil designs gained wide popularity.

First flown in 1935, the German-built **Bücker Bü 133 Jungmeister** is typical of this type of aircraft. A rugged biplane powered originally by a single 160-horsepower Siemens Sh.14 radial, the Bü 133A was designed as a two-seat advanced trainer. Highly maneuverable with a high power-to-weight ratio and fitted with ailerons on both sets of wings, the Jungmeister (Young Master) quickly gained a loyal following among aerobatic contestants.

NASM's Jungmeister is the most famous of all. In 1936, Romanian aerobatic champion Alex Papana brought this aircraft to the United States on board the German airship *Hindenburg*. One year later, after competing in the National Air Races in Cleveland, and flying throughout the United States for several years, Papana sold the aircraft to American Mike Murphy in 1939. The aircraft had been severely damaged when a P-12 landed on top of it at Midway Airport, Chicago. Murphy repaired the Jungmeister and flew it to victory in the U.S. Aerobatic championships before selling it in 1941 to another former U.S. Aerobatic Champion, Beverly "Bevo" Howard.

After fitting a new, more-powerful 185-horsepower Warner Scarab engine, Howard flew his Bü 133 for many years in air shows and competitions, particularly at the graduation ceremonies of the Hawthorne Flying Service of Charleston, South Carolina, of which he was president. Regrettably, Howard lost his life in this aircraft during a public performance in Greenville, North Carolina, in October 1971. His estate honored his wishes and donated the restored aircraft to NASM in 1973.

Art Scholl, another famous aerobatic pilot, is also well represented in the collection. His specially modified **de Havilland DHC-1A Super Chipmunk**, which he flew in international exhibitions for several years, was given to the Museum in 1987. In mid-1945, de Havilland Aircraft of Canada had decided to begin development of a tandem two-seat trainer as a follow-on to the British parent-company's long line of distinguished trainers. First flown on May 22, 1946, the DHC-1 Chipmunk was the first original design produced by the de Havilland Aircraft of Canada and it found a ready market throughout the British Commonwealth. The all-metal monoplane featured a single-spar wing that was stressed to nine gs. Interestingly, although a Canadian design, the Chipmunk was built in greater numbers in Great Britain by the parent company. Of the total 1,219 built, only 219 were produced in Canada. The British version was stressed to 10 gs and featured a framed canopy and minor variations that included a more-forward-mounted landing gear.

Famed aerobatic pilot Art Scholl flew this modified de Havilland DHC-1A Super Chipmunk in breathtaking aerial performances at numerous air-shows in the 1960s, 70s, and 80s.

Leo Loudenslager won seven U.S. and one world aerobatics championship in his Laser 200.

Art Scholl acquired his Canadian-built Chipmunk in 1968 and promptly made several substantial modifications. While keeping the aircraft's bubble canopy, Scholl converted the aircraft to a single-seater, clipped the wings by 18 inches, installed a new fin and rudder, and fitted a retractable landing gear. In addition, he removed the flaps and extended the ailerons. Scholl greatly increased its performance by exchanging the standard 140-horsepower Gypsy Major engine with a 280-horsepower Lycoming GO-480-G1D6 that drove a variable-pitch propeller. This Super Chipmunk was now capable of astounding aerobatics in the talented hands of Art Scholl, who flew it in countless breathtaking aerial demonstrations for many years. Only his untimely death while filming scenes in the movie "Top Gun" cut short his promising career.

During the Second World War, Homestead, Florida, resident Curtis Pitts envisioned a tiny biplane powered by a single Lycoming engine ranging in power from 120-150 horsepower and possessing much greater aerial maneuverability than larger contemporary types. His first version crashed in 1945 after flying for only two weeks. Undaunted, Pitts persisted and in 1946 finished his second aircraft and the first production version of the soon-to-be-famous Pitts Special S-1C. Flown in air shows for several years by Phil Quigley, who was in Pitts's employ as test pilot, the Special was sold to Betty Skelton in 1948 for $3,000. She named the diminutive aircraft *Little Stinker Too* in deference to her Great Lakes 2T1A in which she won her first Feminine International Aerobatic Championship in January 1948.

In this aircraft, Skelton won her second championship in 1949 after modifying the aircraft with a different propeller and installing a second turn-and-bank indicator for inverted flight. Repainting the aircraft in its distinctive red and white stripes, Skelton renamed the aircraft *Little Stinker* and flew it to victory in the 1950 Feminine International Aerobatic Championship. Betty Skelton later sold the aircraft but eventually reacquired her old friend. In 1985, she and her husband Donald Frankman donated it to the Museum.

But this is not the only Pitts Special in the collection. Based partly on the success of *Little Stinker* and numerous impressive aerobatic displays later by other pilots in their own Specials, Pitts's design slowly gained a loyal following in the aerobatic community. The aircraft became a big success in the early 1960s after Curtis Pitts made the aircraft plans available to the homebuilder market. The Pitts Special soon became the aircraft of choice in U.S. aerobatic competitions with more than 250 built. A Pitts Special won the U.S. National Aerobatic Championship in 1966 and for many years thereafter until it was supplanted by monoplane designs.

In 1973, the Smithsonian borrowed and later acquired the S-1S Special *Maryann* from U.S. National Aerobatic team member J. Dawson Ransome to commemorate the team's victory in the 1972 World Championships. This slightly larger version of the popular S-1C featured four ailerons rather than the original two symmetrical airfoils, and a 180-horsepower Lycoming power plant.

With the **Laser 200**, Leo Loudenslager won an unprecedented seven U.S. National Aerobatic Championship titles between 1975 and 1982, as well as the 1980 World Champion title. The airplane originated as a Stephens Akro, but by 1975 Loudenslager had completely modified the airplane with a new forward fuselage, wings, tail, and cockpit. The Laser 200 emerged as a lighter, stronger, and more powerful aircraft, enabling Loudenslager to perform sharper and more difficult maneuvers.

In 1971, Loudenlagser competed in his first contest at the second level of competition and then immediately proceeded to the highest level, unlimited. He flew at the U.S. nationals and amazingly came in eighth in the men's division. He won his first U.S. National Championship title in 1975 and repeated in 1976, 1977, and 1978, and then again in 1980, 1981, and 1982. Aerobatic champion and judge Clint McHenry once said he had only seen two perfect aerobatic routines and both were flown by Loudenslager. Retiring from competition flight in 1983, Loudenslager continued to fly at air shows around the country until his death (not flight-related) in 1997. In 1983, the Laser was painted in the brilliant red Bud Light scheme to reflect its sponsorship.

Loudenslager's legacy is evident in the tumbling and twisting but precise routines flown by current champions and air show pilots. The Laser 200 heavily influenced the look and performance of the next generation of aerobatic aircraft, including the Extra, which dominated competition throughout the 1990s. The Laser 200 was donated by the Loudenslager family in 1999.

The Museum's Sukhoi Su-26M was flown by the Soviet National Aerobatics Team in 1990 and 1991. The Su-26 is known for its outstanding strength and maneuverability.

For decades international aerobatic competitions were dominated by competitors from Eastern Europe and the former Soviet Union. In 1983, the Sukhoi Design Bureau, makers of Soviet and Russian military aircraft, designed the Su-26 for unlimited aerobatic competition. First flown in June 1984, the **Su-26M** was first used by the Soviet Aerobatic Team in August 1984 to participate in the World Aerobatic Championships in Hungary. The Su-26 showed its mettle during the next world competition, which was

held in the United Kingdom in 1986, when the Soviet team won the coveted team prize for the men's competition. The Su-26M airplane routinely challenged the German Extra and French Cap series aerobatic and air show airplanes in national as well as world competitions. Soviet and later Russian pilots flew the design to multiple aerobatic titles. Since the early 1990s, the design has been a very successful export for Russia and is now used by many European and American aerobatic competitors and air show pilots.

The Su-26M is made of more than 50 percent composite materials and has a special symmetrical wing section and arched, cantilever, titanium landing gear. It was designed to handle loads from +12 gs to -10 gs and is a superior aircraft for the most highly skilled pilot. It can perform spectacular gyroscopic maneuvers and quick, multiple snap-rolls, and can nearly hover from its propeller.

The Museum's Su-26M, built in 1990, was flown by the Soviet National Aerobatic team in 1990 and 1991. U.S. aerobatic pilot Gerry Molidor bought it in 1998 and flew it in advanced and unlimited competition until September 2001, when he was forced to retire from unlimited aerobatic competition due to a serious eye injury. The airplane was subsequently acquired by Anheuser-Busch, which donated it to the Museum in 2003.

While not involved in competitive aerobatics, the art of skywriting combines aerobatics with advertising. The TravelAir D4D is one of more than 1,200 Travel Air open-cockpit biplanes built between 1925 and 1930. Popular and rugged, Travel Airs earned their keep as utility workhorses and record breakers. The design was the first success for three giants of the general aviation industry, Lloyd Stearman, Walter Beech, and Clyde Cessna, who in 1925 established the Travel Air Manufacturing Company in Wichita, Kansas.

From 1931 to 1953, Andy Stinis performed skywriting in this airplane for Pepsi-Cola. During those years, skywriting with smoke was a premier form of advertising, and Pepsi-Cola used it more than any other company. Pepsi-Cola acquired the airplane in 1973 and used it for air show and advertising duty until retiring it in 2000. Peggy Davies and Suzanne Oliver, the world's only active female skywriters since 1977, performed in it until the aircraft was grounded. The Pepsi-Cola Company donated the aircraft in 2000.

In 2000, NASM acquired a truly legendary aerobatic aircraft from a truly legendary pilot. Flown by R.A. "Bob" Hoover for nearly 20 years, N500RA is the most recognized **Shrike Commander** in the world. Hoover used his extensive test pilot and fighter pilot skills to become a legendary air show pilot and brought a simple business aircraft to international attention.

The Shrike Commander is a descendent of the 1948 L-3805 Aero Commander, a light, twin-engined, six/seven-seat, high-wing aircraft built by the Aero Design and Engineering Corporation of Culver City, California. In 1960, the company changed its name to Rockwell Standard and then merged with North American in 1967. The aircraft became the Shrike Commander, a twin-engine business aircraft powered by two fuel-injected, 290-horse-power, piston-turbo Lycoming engines. In 1973, the Rockwell International name appeared as a result of more mergers. Between 1968 and 1979, when production ended, Rockwell produced 316 Shrike Commanders.

In the 1950s, Bob Hoover, a World War II combat fighter pilot and accomplished test pilot who flew the chase plane on Chuck Yeager's historic first supersonic flight in 1947, began flying North American aircraft, most famously the P-51 Mustang, at military bases and then at major civilian air shows. In 1973, he began flying the Shrike Commander model in civilian air shows across the country. Hoover bought the aircraft in 1979 and painted it in a distinctive white and green paint scheme with his name on the top of each wing. Hoover's flight routine demonstrated the Shrike's excellent high- and low-speed handling capabilities and its one-engine and no-engine performance. But because it was a stock business aircraft that lacked highly modified engines and quick climb or turn characteristics, the Shrike Commander was a more challenging aircraft for air show flight than his P-51 fighter. In addition to 16-point rolls and loops, Hoover flew a precise deadstick (no-engine) maneuver with a loop, eight-point roll, a 180-degree turn to a touchdown with first one wheel and then the other wheel, landing, and taxi to air show center.

Bob Hoover performed aerial acrobatics for 20 years in his Shrike Commander, a twin-engined business aircraft.

One particular maneuver demonstrated Hoover's superb pilot skills in both the Shrike and the Sabreliner, but it is only visible on film. At altitude, Hoover set a glass on top of the instrument panel and proceeded to pour iced tea into the glass from a pitcher in his right hand while using his left hand to completely roll the aircraft. Combining centrifugal force with smooth handling of the controls, he never spilled a drop of tea. In 2000, Hoover decided to retire the aircraft and gave it to the Museum, flying it to Dulles himself and literally taxiing it into the building upon arrival.

NASM has many other interesting and unique aircraft ranging from high-performance racing machines to ultralights and homebuilts, and even to flying automobiles. These, too, form an important part of the National Aeronautical Collection.

Forced to suspend competition until after the war, the National Air Races resumed competition in 1946. Because surplus high-performance military aircraft were available and because of the demise of the private aircraft builder, the next generation of air race competitors flew modified fighters, particularly P-51s and P-39s. One of these individuals was famed pilot Paul Mantz, who had established a career as a racer and as a stunt pilot for the movies.

Mantz acquired a surplus **North American P-51C Mustang** in the hope of competing in the reborn Bendix transcontinental air race. Mantz greatly improved the Mustang's already excellent range by sealing the interior of the wings and using this space as a fuel tank. This "wet wing" enabled Mantz to fly non-stop from Los Angeles to Cleveland, the home of the National Air Races, and win the 1946 and 1947 Bendix Trophy. Naming his Mustang *Blaze of Noon*, Mantz piloted his aircraft to a coast-to-coast speed record in both directions. This same aircraft, with other pilots at the controls, placed second in 1948 and third in 1949, a remarkable achievement.

Charlie Blair flew Excalibur III*, a modified North American P-51C Mustang, on a record-setting flight across the North Pole in 1951.*

After the 1949 Bendix, Mantz sold his P-51C to Pan American pilot Charles F. Blair, who hoped to set a solo, around-the world speed record. Blair, who had set several flying boat records while flying for American Export

Bournemouth, England, in just under 77 hours (February 8–11, 2006; 25,766 miles). Finally, he set a closed-course distance record over Salina, Kansas, of 74.5 hours (March 14–17, 2006; 25,506 miles).

The all-composite *Global Flyer* was designed and built by Scaled Composites under the direction of Burt Rutan. It contains 13 tanks that hold 2,915 gallons of fuel. Eighty-three percent of the maximum weight is fuel. Fossett delivered the aircraft to NASM in person, flying it into Washington Dulles Airport to a delighted crowd in the summer of 2006.

Though not pure sailplanes, motorized gliders combine the performance of a sailplane with the convenience and safety of an engine. In 1946, Ted Nelson hired noted designer Hawley Harvey Bowlus to create a motor-glider. Based heavily on Bowlus's *Albatross* sailplane, the Nelson BB-1 Dragonfly featured a glide ratio of 20:1 and a single 25-horsepower, two-stroke, four-cylinder Nelson H-29 engine. The first motor-glider ever to receive a certificate of airworthiness, the Dragonfly was unfortunately underpowered and, at $3,000 a copy, too expensive for the market.

Two years later, Nelson produced his improved PG-185-B Hummingbird, which incorporated a 40-horsepower engine. The prototype, which is in NASM, was built of wood but the six production versions were of all metal construction. Mr. Charles Rhoades gave both the Dragonfly and the Hummingbird to the Museum in 1973.

"Sport aviation," "flying for fun," and a host of other labels ably describe the many facets of the leisure side of general aviation. Sailplanes and motorized gliders comprise the more expensive branch of activity while hang gliders and ultralights, which are effectively powered hang gliders, bring affordable flight to a much wider audience than previously possible. NASM possesses an Eipper-Formance 10 and a Valkyrie hang glider, both from 1975, and a **Mitchell U-2 Superwing** ultralight that started its life as a hang glider. The unique Ultraflight Lazaire SS EZ is an advanced twin-engined ultralight with an enclosed fuselage and an inverted "V" tail. The Museum's example flew with and was donated by the police department of Monterey Park, California, where it performed aerial surveillance at a much lower cost than conventional police helicopters.

The Mitchell U-2 Superwing installed in the Museum.

The unique Pathfinder Plus *is an unmanned, solar-powered aircraft that has set several altitude records.*

In the late 1970s, others sought to build the ultimate flying machine, a human-powered aircraft. Dr. Paul MacCready and his close-knit team of engineers, friends, and family members turned this dream into a reality on August 23, 1977, when Bryan Allen piloted and pedaled the MacCready *Gossamer Condor* around a one-half-mile-long figure-eight course, flying above a 10-foot-high obstacle at each end. This triumph won the £50,000 Kremer Prize for the first successful human-powered aircraft and inspired Dr. MacCready to even loftier goals.

After donating the Condor to the Museum in 1978, where it remains on display downtown, Dr. MacCready sought the same challenge that confronted Louis Blériot 70 years before—to fly across the English Channel. Competing for a new Kremer Prize of more than £100,000, MacCready and company fashioned what was essentially a stronger Condor that could be easily transported. Built of carbon fiber-reinforced plastic tubing instead of aluminum, and covered by Mylar plastic, the new *Gossamer Albatross* was a high-wing monoplane with a canard elevator that weighed less and was stronger than the original *Gossamer Condor*. Propulsion was provided by the pilot, who turned a two-bladed propeller by pedaling from a reclined

position inside the fuselage. On June 12, 1979, the *Gossamer Albatross*, again with Bryan Allen as pilot, became the first human-powered aircraft to fly across the English Channel. The flight lasted 2 hours and 49 minutes and covered 22.5 miles between Folkestone, England, and Cap Gris Nez, France. Despite calm winds, the flight took much longer than expected and brought Bryan Allen to the verge of exhaustion.

This record stood for years until a team from the Massachusetts Institute of Technology under the leadership of John Langford created the *Daedalus*, a sophisticated human-powered aircraft designed to replicate the mythical journey of Daedalus from Crete to the Greek mainland. Built of carbon epoxy and covered in Mylar, and with 14-time Greek national bicycle champion Kanelos Kanellopolous at the controls, *Daedalus* flew from Crete to the island of Santorini, a distance of 72 miles, in just under four hours. An unexpected gust of wind caused the craft to break up and gently crash into the Aegean Sea but not before the record had been set. NASM owns the only surviving pieces.

Inspired by the success of the *Gossamer Albatross*, Dr. MacCready and his team from AeroVironment turned their attention to the design and construction of a host of radical high-lift, low-weight aircraft powered by unconventional means. MacCready sought to combine his unique aircraft technology with solar power, and his first success came with the *Gossamer Penguin*, which made its first flight solely on solar cell power on May 18, 1980. The flight lasted only about 30 seconds, with Marshall MacCready as pilot.

Spurred by this achievement, the design team proceeded to build a long-endurance solar-powered aircraft. After much work, the result was the *Solar Challenger*, a 217-pound aircraft with a wingspan of 47 feet and a length of 30 feet. Power is provided by 16,128 photovoltaic cells that cover the wing and the horizontal stabilizer and produce a steady 2.7 horsepower from its electric motor. As long as the sun shines, the *Solar Challenger* has an unlimited endurance. To demonstrate the possibilities of solar power and energy conservation, MacCready sent the aircraft up in a series of proving flights. The most dramatic occurred on July 7, 1981, when pilot Stephen Ptacek flew the aircraft across the English Channel from Cormeillesen-Vexin, France, to the Royal Air Force Base at Manston, England. The 5-hour and 23-minute flight covered a distance of 230 miles during which Ptacek reached a top speed of 47 miles per hour and an altitude of 12,000 feet. This success has led AeroVironment to produce exotic experimental high-altitude, unmanned reconnaissance aircraft that typify MacCready's innovative ideas and his special knack for problem-solving. The following year, Dr. MacCready presented the aircraft to the Museum, where it was briefly on display before it was lent to the Science Museum of Virginia in Richmond.

Perhaps the most extreme iteration of MacCready's ultra-lightweight solar-powered design is the ***Pathfinder Plus*** unmanned aircraft. Designed and built by AeroVironment, Pathfinder Plus is a high-altitude, solar-powered,

unmanned experimental aircraft intended to explore the possibilities of unlimited-duration, high-altitude reconnaissance. During the 1990s, it conducted 10 test flights, three of which set altitude records, the highest of which was 24,445 meters (80,201 feet). The aircraft was built under the sponsorship of the Lawrence Livermore National Laboratory and the Ballistic Missile Defense Organization. It has a carbon-fiber main spar and is covered with a polymer skin and silicon solar cells that power eight electric motors. The project was later managed by NASA's Dryden Flight Research Center before the machine was given to the Museum by its builder in 2007.

The graceful Lockheed L-1049 Super Constellation set the standard for air travel in the 1950s flying with the airlines and the military.

On a much larger scale than general aviation, civil aviation embraces one of the world's great industries—commercial air transportation. The major technological developments in this field have had a profound effect not only on the airlines and their rise to prominence but on the way that we now

view our world. In the 1920s, airlines linked countries in hours rather than in days spent in surface travel. The development of large four-engined, piston-powered airliners of the 1940s and 1950s first united the continents in hours rather than days spent in the elegant but slow flying-boats of the 1930s. By the late 1950s, jet airliners made it possible to travel virtually anywhere on the planet in a day. Such an achievement, the social, political, and economic effects of which have helped to create global economies, is reflected in some of the major exhibits and artifacts in the National Air and Space Museum.

Air travel in the 1940s and 1950s was the heyday of the large piston-engined airliner. Air fares were still beyond the reach of most people and air travel was confined to luxury travel for the wealthy, the privileged, and the business executive. Air travel was elegant and travelers dressed appropriately. One aircraft that well illustrates that romantic age is the graceful, dolphin-shaped Lockheed Constellation.

The introduction of the Lockheed Constellation into commercial service following the end of the Second World War greatly intensified transcontinental competition. Sleek and powerful, with their distinctive triple fins and rudders (designed so that the aircraft could fit inside existing hangars), the graceful "Connies" dramatically lowered travel times and introduced unheard-of pressurized comfort. When Transcontinental and Western Air (TWA) began flying Constellations in March 1946, passengers could travel from New York to Los Angeles in 11 hours with only one 25-minute fuel stop, five hours shorter than United's New York–San Francisco schedule with the slower, unpressurized Douglas DC-4s. The one-way fare of $118.30 matched United's, although TWA added an extra $25 for the Constellation's "Advanced Sky Chief" service. Connies became so successful that by the 1950s hundreds were in service with airlines around the world providing unprecedented levels of high-quality air service to thousands of travelers each day.

By the early 1950s in the United States, competition between manufacturers and between airlines was intense. The pressurized DC-6 and later the longer-ranged Douglas DC-7 transports forced Lockheed to develop improved versions of the standard Constellation series. In 1951, the **Lockheed L-1049 Super Constellation** entered service with Eastern Air Lines. This was a "stretched" version of the standard 44-seat Constellation and could carry 71 passengers. The L-1049C and later variants were powered by four new Wright R-3350 Turbo Compound engines that initially produced 3,250 horsepower, giving the aircraft a top speed in excess of 350 miles per hour. Northwest Orient and TWA in particular also operated Super Constellations. These aircraft and their competitors from Douglas ruled the skies until the advent of jet travel.

The L-1049F variant of the Super Constellation, featuring a strengthened landing gear, was used by the U.S. Air Force to fly cargo and personnel for many years. After being withdrawn from frontline duty, some of the 33 L-1049Cs flew with all-cargo airlines as well as second- and third-level airlines around the world. The Museum's Super Constellation is one of these.

Acquired in 1988, our L-1049C started life as an Air Force C-121C flying for the 1608th Air Transport Wing of the Atlantic Division of the Military Air Transport Service based in Charleston, South Carolina. For more than 20 years, this Connie flew cargo and personnel around the globe, later being used by the Mississippi, West Virginia, and Pennsylvania Air National Guards before it was sold to a civilian operator. When plans to create a small airline serving Las Vegas and Los Angeles fell through, the aircraft was acquired by NASM. It was flown across the country to Washington Dulles International Airport where it resides today, restored in its markings of the 167th Military Airlift Squadron, West Virginia Air National Guard.

Immediately after the Second World War, airlines began to search for a more efficient, pressurized replacement for the venerable Douglas DC-3. Engineers at Boeing and Lockheed offered proposals but none saw production. Designers at Martin and Convair, however, did find success. The Martin 2-0-2 and 4-0-4 series were widely flown but their careers were marred by structural problems. Far more successful was the pressurized Convair 240 series.

Searching for a replacement for the DC-3 in 1945, American Airlines initiated a design requirement for a modern, pressurized, twin-engined, 40-seat airliner for use in local service. The product was the Convair 240 (with 2 engines and 40 passengers—hence the name). As launch customer, American placed an initial order for 100 of the new CV-240s (later reduced to 75).

The CV-240 first flew on March 16, 1947, and was delivered to American on February 28, 1948. It was the first pressurized twin-engined airliner to enter service. A more powerful version built for United Air Lines was introduced in 1952 as the Convair 340. This aircraft featured a longer fuselage to carry 44 passengers as well as a larger wing that gave the aircraft improved high-altitude performance. The last piston-engined variant was the CV-440 Metropolitan, which incorporated improved sound-proofing and optional weather-radar. Some 170 340/440s were eventually converted into CV 580s by replacing the Pratt & Whitney R-2800 piston engines with Allison 501-D13 turboprops of 3,750 shaft horsepower. Sales of the 240/340/440 series to the airlines and the military totaled more than 1,000 between 1947 and 1956.

The Museum is fortunate to preserve in its collection a very special Convair 240. This is the *Caroline*, the first private aircraft ever used by a candidate during a presidential campaign. It was used by Sen. John F. Kennedy during his successful 1960 campaign for the Democratic Party nomination and subsequent campaign for president. Historians credit this aircraft with providing Kennedy with the narrow margin of victory because it allowed him to campaign more effectively during that very hotly contested race. The *Caroline*, named after President Kennedy's daughter, revolutionized American politics; since 1960, all presidential candidates have used aircraft as their primary means of transportation.

This historic aircraft was built in August 1948 and delivered to American Airlines, which operated it until 1959 when Joseph Kennedy purchased the aircraft and fitted it with an executive interior for his son's 1960 presidential campaign. For security reasons, President Kennedy rarely used the aircraft after the election but the family used it until 1967.

In September 1967, on behalf of his family, Sen. Edward Kennedy approached Smithsonian Secretary S. Dillon Ripley with an offer to donate the aircraft. Recognizing the eventual historical significant of the aircraft, Ripley suggested that NASM accept the aircraft. The aircraft was donated on November 17, 1967, during a ceremony at National Airport attended by Ripley and the Kennedy family. Interestingly, from 1975 until 1982, the Smithsonian Museum of American History recognized the significance of the *Caroline* by borrowing the interior for its popular "We the People" exhibit. Following the close of the exhibit, the interior was lent to the Kennedy Library until it was returned in 1993. It is currently in storage.

A revolution occurred in air travel when, in the 1950s, powerful and durable jets enabled aircraft manufacturers to build bigger, faster, and more-productive airliners. With their high speeds and low operating costs, jet airliners enabled the airlines to provide lower fares, which greatly expanded the passenger demand for air travel.

Although jet airliners such as the British de Havilland D.H. 106 Comet and the Soviet Tupolev Tu-104 entered service earlier, the Boeing 707 and Douglas DC-8 were bigger, faster, had greater range, and were more profitable to fly. First flown in December 1957, the 707 was developed from the 367-80, or "Dash 80," the prototype for the U.S. Air Force's KC-135 jet tanker and the progenitor of Boeing's long line of distinguished jet transports.

In the early 1950s, Boeing had begun to study the possibility of creating a jet-powered military transport and tanker to complement the new generation of Boeing jet bombers entering service with the U.S. Air Force. When the Air Force showed no interest, Boeing invested $16 million of its own capital to build a prototype jet transport in a daring gamble that the airlines and the Air Force would buy it once the aircraft had flown and proven itself. As Boeing had done with the B-17, it risked the company on one roll of the dice and won—handsomely.

Boeing engineers had initially based the jet transport on studies of improved designs of the Boeing 367-80, better known to the public as the C-97 Stratofreighter, a piston-engined transport and aerial tanker based on the B-29 bomber. By the time Boeing progressed to the 80th iteration, the design bore no resemblance to the C-97 but, for security reasons, Boeing decided to let the jet project be known as the 367-80, or the **Dash 80**.

Work proceeded quickly after the formal start of the project on May 20, 1952. The 367-80 mated a large cabin based on the dimensions of the C-97 with the 35-degree swept-wing design based on the wings of the B-47 and B-52 but considerably stiffer and incorporating a pronounced dihedral. The wings were mounted low on the fuselage and incorporated high-speed and low-speed ailerons as well as a sophisticated flap and spoiler system. Four Pratt & Whitney JT3 turbojet engines, each producing 10,000 pounds of thrust, were mounted on struts beneath the wings.

Upon the Dash 80's first flight on July 15, 1954 (the 34th anniversary of the founding of the Boeing Company), Boeing clearly had a winner. Significantly larger than the de Havilland Comet and flying 100 miles per hour faster, the new Boeing had a maximum range of more than 3,500 miles. As hoped, the Air Force bought the design as a tanker/transport, assigning it the designation of KC-135A.

Boeing quickly turned its attention to selling the airline industry on this new jet transport. Clearly the industry was impressed with the capabilities of the prototype 707 but never more so than at the August 1955 Gold Cup hydroplane races held on Lake Washington in Seattle. During the festivities surrounding this event, Boeing had gathered many airline representatives to enjoy the competition and witness a flypast of the new Dash 80. To the audience's intense delight and Boeing's profound shock, test pilot Alvin "Tex" Johnson barrel-rolled the Dash 80 over the lake in full view of thousands of astonished spectators. While breaking every rule imaginable, Johnson's vividly display convinced the airline industry of the Dash 80's superior strength and performance.

below and opposite:
The Boeing 367-80, or Dash 80, was the prototype for America's first jet airliner—the 707.

In searching for a market, Boeing found a ready customer in Pan American Airway's president Juan Trippe, who had been spending much of his time searching for a suitable jet airliner to enable his pioneering company to maintain its leadership in international air travel. Working with Boeing, Trippe overcame Boeing's resistance to widening the Dash-80 design, now known as the 707, to seat six passengers in each seat row rather than five. Trippe did so by placing an order with Boeing for 20 707s but also ordering 25 of Douglas's competing DC-8, which had yet to fly but could accommodate six-abreast seating. At Pan Am's insistence, the 707 was made four inches wider than the Dash 80 so that it could carry 160 passengers six-abreast.

In October 1958, Pan American ushered the jet age into the United States when it opened international service with the Boeing 707. National Airlines inaugurated domestic jet service two months later using a 707-120 borrowed from Pan Am. American Airlines flew the first domestic 707 jet service with its own aircraft in January 1959, and set a new speed mark when it opened the first regularly-scheduled transcontinental jet service later that year. Subsequent nonstop flights between New York and San Francisco took only five hours, three hours less than by the piston-engine DC-7. The one-way fare, including a $10 surcharge for jet service, was $115.50, or $231 for a round trip. The flight was almost 40 percent faster and almost 25 percent cheaper than flying by piston-engine airliners. The consequent surge of traffic demand was substantial.

The 707 was originally designed for transcontinental or one-stop transatlantic range. But modified with extra fuel tanks and more efficient turbofan engines, the 707-300 Intercontinental series aircraft could fly nonstop across the Atlantic with full payload under any conditions. Boeing built 855 707s, of which 725 were bought by airlines worldwide.

The Dash 80's career did not end after the first flight of the 707. This remarkable aircraft served as a test bed for numerous Boeing and NASA projects for the next 15 years until it was retired to NASM in 1972. It tested new engines for the 707, different wings, flaps, airfoil shapes, and a variety of other important features, many of which were incorporated into later aircraft.

In the 1950s, air travel was revolutionized with the advent of jet propulsion. First the de Havilland Comet and, later, the Boeing 707, greatly increased the speed of travel from 350 to over 600 miles per hour. Airlines and customers flocked to the new jet airliners as travel times were cut dramatically and the seat-mile costs to the airlines dropped. The conclusion drawn by engineers, managers, and politicians seemed clear: the faster the better. Unfortunately, they were wrong.

In Europe, enterprising designers in Great Britain and France were independently outlining their plans for a supersonic transport (SST). In November 1962, in a move reminiscent of the Entente Cordiale of 1904, the two nations agreed to pool their resources and share the risks in building this new aircraft. They also hoped to highlight Europe's growing economic unity as well as its aerospace expertise in a dramatic and risky bid to supplant

following spread, clockwise from bottom:
The Concorde was the world's only supersonic airliner to attain sustained service, flying for 27 years before the fleet was retired in 2003.

Because of the high angle-of-attack required by delta wings, the Concorde featured exceptionally tall landing gear.

Four massive Bristol/SNECMA Olympus turbofans powered the Concorde.

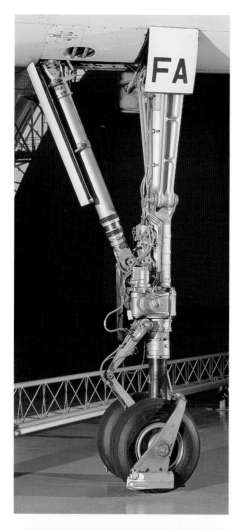

the United States as the leader in commercial aviation. The aircraft's name reflected the shared hopes of each nation for success through cooperation—**Concorde**.

The designers at the British Aircraft Corporation and Sud Aviation, later reorganized as Aerospatiale, quickly settled on a slim, graceful form featuring an ogival delta wing that possessed excellent low-speed and high-speed handling characteristics. Power was to be provided by four massive Olympus turbofan engines built by Rolls-Royce and SNECMA. Realizing that this first-generation SST would cater to wealthier passenger, Concorde's designers created an aircraft that carried only 100 seats in tight, four-across rows. They assumed that first-class passengers would flock to the Concorde to save valuable time while economy-class passengers would remain in the larger, slower subsonic airliners.

While mounting costs constantly threatened the program, as construction continued with exactly 50 percent of each aircraft in each country, the first Concorde was ready for flight in 1969. With famed French test pilot Andre Turcot at the controls, Concorde 001, which was assembled at Toulouse, took to the air on March 2, 1969. Although the Soviets had flown their version of the SST first, the Tupolev Tu-144 had been rushed into production and suffered from technological problems that could never be solved. Following the successful first flight, a total of four prototype and preproduction Concordes were built and thoroughly tested and by 1976 the first production Concordes were ready for service.

But across the Atlantic all was not rosy. During this time, America sought to produce its own bigger and faster SST. After a contentious political debate, the federal government refused to back the project in 1971, citing

environmental problems, particularly noise—the sonic boom—and engine emissions that were thought to harm the upper atmosphere. Anti-SST political activity in the United States delayed the granting of landing rights, particularly into New York City, causing further delays.

More ominously for the Concorde, no airlines placed orders for this advanced SST. Despite initial enthusiasm, the airlines dropped their purchase options once they calculated the operating costs of the Concorde. Consequently, only Air France and British Airways—the national airlines of their respective countries—flew the 16 production aircraft, and only after purchasing them from their governments at virtually no cost.

Nevertheless, in January 1976, Concorde service began and by November these graceful SSTs were flying to the United States. A technological masterpiece, each Concorde smoothly transitioned to supersonic flight with no discernable disturbance to the passenger. In service, the Concorde would cruise at twice the speed of sound between 55,000 and 60,000 feet—so high that passengers could actually see the curvature of the Earth. The Concorde was so fast that, despite outside temperatures lower than −56 degrees Celsius, the aircraft's aluminum skin would heat up to over 120 degrees Celsius while the Concorde actually expanded eight inches in length. The interior of the window gradually grew quite warm to the touch. And all the while, each passenger was carefully attended to while enjoying a magnificent meal and superb service. Transatlantic flight time was cut in half with the average flight taking less than four hours.

For the next 27 years, supersonic travel was the norm for the world's business and entertainment elite. But eventually the harsh reality of the economic marketplace forced Air France and British Airways to cut back their already

The SAAC-23 Learjet created the market for small, jet-powered private aircraft.

limited service. Routes from London and Paris to Washington, Rio de Janeiro, Caracas, Miami, Singapore, and other locations were cut, leaving only the transatlantic service to New York. And even on most of these flights, the Concorde flew half full with many of the passengers flying as guests of the airlines or as upgrades. With the average round-trip ticket costing more than $12,000, few could afford to fly this magnificent aircraft. Operating costs escalated as parts became more difficult to acquire and, with an average of one ton of fuel consumed per seat, the already small market for the Concorde gradually grew smaller.

Despite the excellence of the Concorde's design, its operators realized that its days were numbered because of its high costs. In 1989, in commemoration of the 200th anniversary of the French Revolution and the 200th anniversary of the ratification of the U.S. Constitution, the French government sent a copy of the Declaration of the Rights of Man to the United States. Appropriately, this famous document was delivered on the Concorde and with it a promise from Air France to give one of these aircraft to the people of the United States, specifically to the Smithsonian Institution's National Air and Space Museum.

Fourteen years later that promise was fulfilled. In April 2003, Air France president Jean Cyril Spinetta informed the Museum that Concorde service would end on May 31 following the decision by the aircraft's manufacturer to stop supporting the fleet. As planned, on June 12 Air France delivered its most treasured Concorde, F-BVFA, to Washington Dulles International Airport on its last supersonic flight for the airline. Onboard were 60 passengers including Gilles de Robien, the French Minister for Capital Works, Transport, Housing, Tourism, and Marine Affairs, Mr. Spinetta, and several past Air France presidents as well as former Concorde pilots and crew members. In a dignified yet bittersweet ceremony, Mr. Spinetta signed over Concorde *Fox Alpha* to the Museum for permanent safekeeping. The brief age of commercial supersonic flight had come to an end.

In the early 1960s, the jet age reached the general aviation market with the arrival of the remarkable Learjet. Aviation innovator and pioneer William P. Lear envisioned a small, jet-powered private aircraft to fill a need he saw for high-speed business travel. With a courageous belief that the market for a business jet would develop and prosper, Lear formed the Swiss-American Aviation Corporation (SAAC) in 1960 in St. Gallen, Switzerland, to design and build the **SAAC-23 Learjet**.

Using the wing from the stillborn Swiss P-16 strike fighter, Lear constructed a graceful aircraft with two rear-mounted General Electric CJ 610 turbojets, each producing 2,850 pounds of thrust. The aircraft has a "T" tail, low-mounted cantilevered wings, and a pressurized, semi-monocoque fuselage. Capable of carrying nine passengers, the Learjet could cruise at 500 miles per hour, significantly faster than conventional business aircraft, and performed as well as many military fighters. The aircraft first flew on October 7, 1963, one year after the entire operation was reorganized and moved from Switzerland to Wichita, Kansas. The second Model 23 was flown by the factory for testing until 1977 when it was donated to the Museum. Other Learjets have set numerous records and production has steadily continued through a series of different models up to this day.

At the same time that William Lear was developing his diminutive jet, other manufacturers sought to garner their share of this rapidly expanding market. In France, Dassault Aviation realized the potential for the executive jet and produced a graceful 10-seat design based on the wing planform of the successful Dassault Mystère series of military fighters. Originally called the Mystère 20, the Falcon proved an immediate success with customers after its first flight on May 4, 1963. Through an arrangement with Pan American World Airways, Falcons were sold in the United States through the Pan American-owned Falcon Jet Corporation. Known in the U.S. as Fan Jet Falcons, this excellent aircraft also served as the foundation for a new industry.

On April 17, 1973, Memphis businessman Fred Smith started a small overnight package delivery service using two **Dassault Falcon 20**s. The first aircraft for the new Federal Express Corporation, N8FE, is now in the Museum's collection. N8FE and the rest of FedEx's Falcon fleet were extensively modified from their original executive configuration. All the seats were removed, the windows blanked out, and a new oversized cargo door was installed on the left side of the aircraft just behind the cockpit. The gross weight was increased from 25,300 to 28,660 pounds. The result was a highly efficient transport that enabled FedEx to prosper and grow until it was superseded—a victim of its own success—by newer and much larger aircraft. Thirty-three Falcon 20s were operated by FedEx before the fleet was sold off in the early 1980s.

Fred Smith created Federal Express using this Dassault Falcon 20 as his company's very first aircraft.

Possessing far less graceful lines than the Falcon or the LearJet but no less successful in its intended line of work is the Soviet Antonov An-2. Designed not for the rarified world of executive travel but the harsh working environment of the former Soviet Union, the An-2 looks ungainly but is a remarkably successful and practical design. First flown in 1947, this huge single-engine biplane, with an all-metal fuselage wider than that of a DC-3, was intended to fly virtually anywhere in any conditions while carrying up to 3,300 pounds of payload or 14 passengers. With its biplane wings and 1,000-horsepower Shvetsov ASh-62 radial engine, the rugged An-2 possesses remarkable short-field performance that gives it unmatched versatility for a myriad of roles from crop duster to airliner, from cargo hauler to water bomber. In 1959, production was transferred to the PZL plant in Poland where it continues to this day well over a half a century after the An-2 first took to the sky. The Poles have also produced the An-3 turboprop-powered version of this venerable aircraft. More than 20,000 have been built in the former Soviet Union, East Germany, Poland, and China. Our An-2 came from Mr. E.J. "Buzz" Gothard, who acquired the aircraft in India before donating it in 1983.

As ugly and as practical as the An-2, the Grumman G-164 Ag-Cat was the first aircraft specifically designed by a major aircraft company for agricultural aviation and is one of the most successful, as well. Following World War II, agricultural aviation rapidly expanded with the growth of food production for the postwar domestic and export markets. The Grumman Aircraft Company saw the need for a special-purpose "duster" design and, after consulting with agricultural pilots and companies around the country, introduced the Ag-Cat in 1957.

After World War II, surplus Stearman military biplane trainers were pressed into duster service to fulfill the urgent need, many of them structurally reinforced and equipped with surplus 450-horsepower Pratt and Whitney radial engines to handle the rigors of the very low altitude, high-g-force crop-dusting maneuvers. The Stearman was a good airplane but it and other civil aircraft were not designed for sustained flying in this type of environment. In 1955, Grumman preliminary design engineers Joe Lippert and Arthur Koch proposed the design for a "purpose built" crop dusting airplane as a means of fulfilling a pressing need by the agricultural community as well as the perceived need for Grumman to diversify its product lines. A major consideration for their rugged and maneuverable biplane duster concept was the availability of thousands of inexpensive war-surplus 220-horsepower Continental radial engines. The aircraft structure would have to be very strong to accommodate the large hopper, payload, and accompanying spray/spreader equipment and sustain safe flight.

The Ag-Cat's first flight, on May 27, 1957, went exceedingly well, with the second prototype following one month later. Three senior crop-dusting pilots were brought in from various parts of the country to test the airplane and, in the summer of 1958, an extended East Coast demonstration tour allowed well over 100 duster pilots to put the two prototypes through their paces. An upsurge of Grumman military orders were going to prevent the production of the Ag-Cat at the Long Island factory, however, so the board of directors subcontracted the entire program to the Schweizer Aircraft Company of Elmira, New York. The first Schweizer-built Ag-Cat flew in 1959.

The Museum's Grumman G-164 Ag-Cat, one of the early license-built production models built by Schweizer, rolled off the factory line on May 2, 1963. As a crop duster, this aircraft accumulated 12,778 flight hours applying seed, fertilizer, pesticides, and herbicides to nearly any crop that is grown in the United States. The airplane was struck from the records on September 30, 1999, when the engine was removed from the airplane.

In 2004, the National Agricultural Aviation Association (NAAA) contacted NASM regarding the potential addition of modern crop duster aircraft to the collection and learned that the Museum was indeed searching for an appropriate agricultural aircraft. The NAAA alerted the agricultural aviation community and, in August 2005, Ralph Holsclaw and Growers Air Service offered to donate their **Grumman G-164A Super Ag-Cat**. Growers delivered the Ag-Cat, completely cleaned of all chemical residues and restored, to the Museum in March 2008.

In the years following the Second World War, civil aviation flourished beyond all expectations and commercial air transportation spread throughout the world. In varying degrees, national airlines improved and enriched the lives of countless millions of people, rich and poor. Their direct or indirect influence brought the population of the planet closer together. But, on the other side of the coin, the huge advancements in aircraft design also foreshadowed the much greater destructive capability of military aviation. Though the threat of worldwide conflict has so far been avoided, regional enmities have prevailed in many corners of the globe. And these have stimulated a demand for military aircraft of ever greater performance and capability.

The Grumman G-164 Ag-Cat and Super Ag-Cat were designed specifically for agricultural aviation.

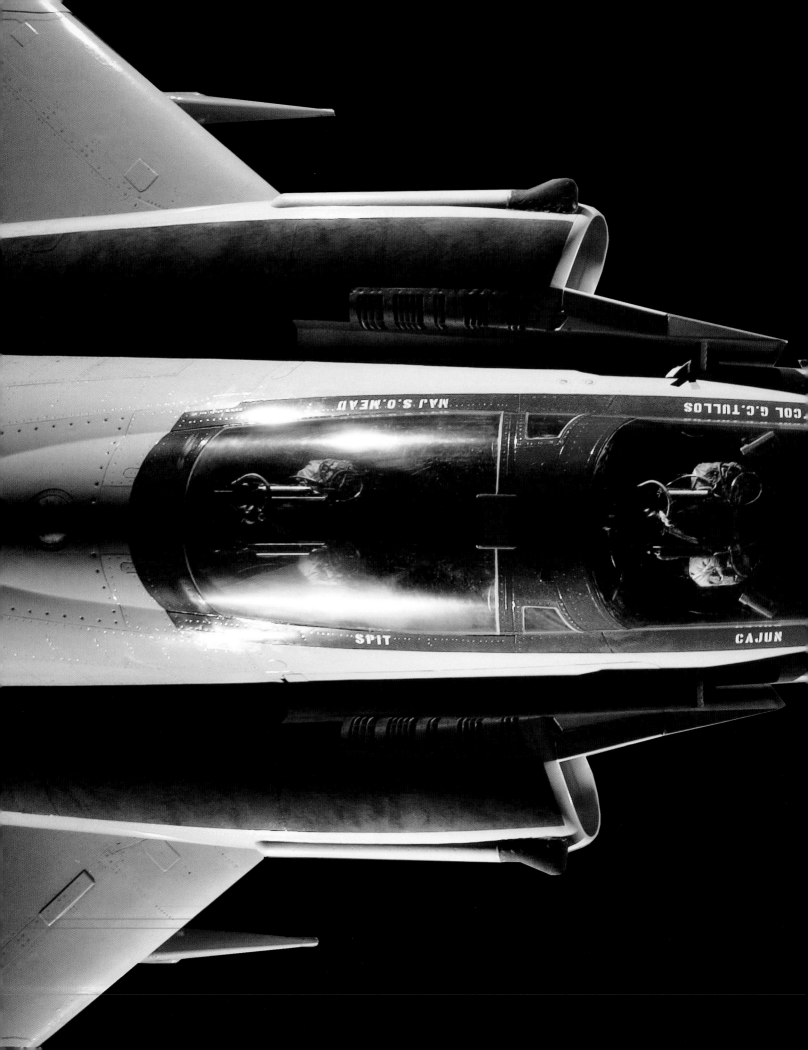

Postwar Military Aviation and Vertical Flight

top and bottom:
With its excellent speed, maneuverability, and firepower, the Soviet-built MiG-15 shocked the West when it entered combat in Korea.

opposite:
Polish Air Force pilot Lt. Franciszek Jarecki wore this leather flightsuit when he defected to the West in his MiG-15.

previous spread:
The McDonnell Douglas F-4S Phantom II is one of the most successful jet fighters in history.

At the end of the Second World War, the United States emerged as an economic, political, and military superpower. Having assembled the largest fleet of military and naval aircraft in history and having used that resource with decisive effect in every theater of war, the United States continued to rely heavily on air power as the primary deterrent to conventional and nuclear conflict during the immediate postwar years. Military aviation had evolved rapidly during the war, and by the end of hostilities, jet-powered aircraft were clearly emerging as the dominant frontline air weapon. New technologies, particularly in vertical flight, promised to revolutionize the battlefield although conventional, piston-engined aircraft still remained important contributors.

Following the lead of Germany and Britain, the United States quickly turned to the development of a new generation of jet-powered aircraft during the mid- to late-1940s. Turbojet engines produced previously unimaginable levels of power that readily lent itself to aircraft propulsion. Because of the high fuel consumption, the first generation of turbojet engines was first incorporated into military fighters, where speed rather than efficiency was the most important performance factor. America's first jet, the Bell XP-59 Airacomet, incorporated a British de Havilland engine designed by Frank Halford and based on Sir Frank Whittle's pioneering centrifugal-flow design. The reliability and adequate power available from this engine guaranteed that the first generation of jet combat engines would use this configuration. Though too slow for frontline service, the XP-59, the first one of which is in the Museum's Milestones of Flight gallery, trained a new generation of U.S. jet fighter pilots who flew the new Lockheed P-80 Shooting Star as the military's first operational jet fighter. The first XP-80 is also in the Museum's collection and is exhibited in the Jet Aviation gallery.

The P-80, which became the F-80 upon creation of the independent U.S. Air Force in September 1947, was a superlative fighter design by Clarence "Kelly" Johnson of the famous Lockheed "Skunk Works." It set numerous speed records after it entered service late in 1945 and proved itself in combat five years later in the skies over Korea, claiming the world's first jet-to-jet victory despite being technologically outclassed by the superlative, second-generation, Soviet swept-wing MiG-15.

In 1947, in response to the high accident rate of the aircraft in service, Lockheed recognized the need for a two-seat trainer version of the P-80. Risking $1 million of company funds, Lockheed assembled a design team under Don Palmer to develop the aircraft. Having received the approval of the Army Air Forces in August 1947, Palmer and his team stretched a P-80C airframe with two fuselage plugs to make room for an additional pilot. A graceful one-piece canopy covered the two pilots and their ejection seats, but fuel capacity was reduced from 425 to 353 gallons. Ironically, the lengthened fuselage was aerodynamically better, thereby giving the new TF-80C slightly improved performance than its F-80C predecessor.